T0280641

SpringerBriefs in Mathematical Physics

Volume 31

SpringerBriefs are characterized in general by their size (50-125 pages) and fast production time (2-3 months compared to 6 months for a monograph).
Briefs are available in print but are intended as a primarily electronic publication to be included in Springer's e-book package.

Typical works might include:

- An extended survey of a field
- A link between new research papers published in journal articles
- A presentation of core concepts that doctoral students must understand in order to make independent contributions
- Lecture notes making a specialist topic accessible for non-specialist readers.

SpringerBriefs in Mathematical Physics showcase, in a compact format, topics of current relevance in the field of mathematical physics. Published titles will encompass all areas of theoretical and mathematical physics. This series is intended for mathematicians, physicists, and other scientists, as well as doctoral students in related areas.

More information about this series at http://www.springer.com/series/11953

A. M. Mathai • H. J. Haubold

Erdélyi–Kober Fractional Calculus

From a Statistical Perspective, Inspired by Solar Neutrino Physics

A. M. Mathai
Department of Mathematics and Statistics
McGill University
Montreal, Canada

H. J. Haubold
Office for Outer Space Affairs
United Nations
Vienna, Austria

Additional material to this book can be downloaded from http://extras.springer.com.

ISSN 2197-1757 ISSN 2197-1765 (electronic)
SpringerBriefs in Mathematical Physics
ISBN 978-981-13-1158-1 ISBN 978-981-13-1159-8 (eBook)
https://doi.org/10.1007/978-981-13-1159-8

Library of Congress Control Number: 2018950187

This Springer imprint is published by the registered company Springer Nature Singapore Pte Ltd.
The registered company address is: 152 Beach Road, #21-01/04 Gateway East, Singapore 189721, Singapore

Preface

This monograph deals with Erdélyi-Kober fractional integrals and fractional derivatives from a statistical perspective, inspired by solar neutrino physics. The application of diffusion entropy analysis to Super-Kamiokande data led to ideas to consider generalizations of entropy (entropic pathway) and diffusion (anomalous diffusion). Examplified by Erdélyi-Kober fractional calculus it is shown that the statistical density of a product of two statistically independently distributed real scalar positive random variables or real positive definite matrix-variate random variables or complex Hermitian positive definite matrix-variate random variables is a constant multiple of the Erdélyi-Kober fractional integral of the second kind of order α and parameter γ, when the density of one of the random variables is arbitrary and the density of the other random variable is a type-1 beta density with the parameters $(\gamma + 1, \alpha)$ in the real scalar case, $(\gamma + \frac{p+1}{2}, \alpha)$ in the real $p \times p$ matrix-variate case, and $(\gamma + p, \alpha)$ in the complex $p \times p$ matrix-variate case. If $x_1 > 0, x_2 > 0$ are the real scalar random variables, $X_1 > O, X_2 > O$ are real $p \times p$ matrix-variate random variables, $\tilde{X}_1 > O, \tilde{X}_2 > O$ are the complex $p \times p$ matrix-variate random variables, then the product is $u_2 = x_1 x_2$ (real scalar case), $U_2 = X_2^{\frac{1}{2}} X_1 X_2^{\frac{1}{2}}$ (real matrix-variate case), and $\tilde{U}_2 = \tilde{X}_2^{\frac{1}{2}} \tilde{X}_1 \tilde{X}_2^{\frac{1}{2}}$ (complex matrix-variate case). Then the density of $u_2, U_2, and \tilde{U}_2$, denoted by $g_2(u_2), g_2(U_2), and \tilde{g}_2(\tilde{U}_2)$, respectively, is a constant multiple of the Erdélyi-Kober fractional integral of the second kind. It is shown that the density of the ratio $u_1 = \frac{x_2}{x_1}, U_1 = X_2^{\frac{1}{2}} X_1^{-1} X_2^{\frac{1}{2}}, and \tilde{U}_1 = \tilde{X}_2^{\frac{1}{2}} \tilde{X}_1^{-1} \tilde{X}_2^{\frac{1}{2}}$, in the real scalar case, in the real $p \times p$ matrix-variate case, and in the complex $p \times p$ matrix-variate case, respectively, is a constant multiple of the Erdélyi-Kober fractional integral of the first kind of order α and parameter γ when one density is arbitrary and the other density is a type-1 beta density with parameters (γ, α) in all the cases. When the functions are not densities, then it is shown that the second kind and first kind Erdélyi-Kober fractional integrals are the Mellin convolution of a product and ratio, respectively, in the scalar case and M-convolution of a product and ratio in the matrix-variate case. General definitions of first kind and second kind Erdélyi-Kober fractional integral operators are established, from where all the various fractional integrals introduced in the literature are available as special cases. These ideas are extended to the real and complex matrix-variate cases. From these fractional integral operators, fractional differential operators are derived, both in the Riemann-Liouville and Caputo senses.

Chapter 1 provides a brief overview on solar neutrino detection and its background in terms of statistical mechanics and neutrino physics. Results of the diffusion entropy analysis of solar neutrino data collected by Super-Kamiokande are provided and discussed in terms of a prospective fractional diffusion model that leads to a diffusion equation in terms of Erdélyi-Kober operators. This result is the basis for the development of Erdélyi-Kober fractional calculus in the following chapters from a statistical perspective.

Chapter 2 covers Erdélyi-Kober fractional integrals in the real scalar variable case. A general notation is introduced to cover all fractional integrals and fractional derivatives. It is shown that all the fractional integrals available in the literature can be obtained as special cases from the general definition given here in terms of statistical densities of product and ratios, Mellin convolutions of products and ratios, or M-convolutions of products and ratios.

Chapter 3 deals with Erdélyi-Kober fractional integrals in the real matrix-variate case. Connections to statistical densities of product and ratio of matrix-variate random variables are also established here.

Chapter 4 introduces Erdélyi-Kober fractional integrals for the real multivariate case. Multivariate means a collection of real scalar variables and real-valued functions of these variables.

Chapter 5 generalizes these scalar variable results to real matrix-variate cases.

Chapter 6 starts with the discussion of Erdélyi-Kober fractional integrals in the complex domain. The necessary tools for handling real-valued scalar functions of matrix argument, when the argument matrix is Hermitian positive definite, are developed in this chapter. Connections to complex matrix-variate statistical distributions are also established. The basic idea in all these developments is statistical densities of products and ratios and their connection to Erdélyi-Kober fractional integrals.

In Chap. 7, differential operators, operating on real and complex matrices, are developed. With the help of these differential operators, fractional derivatives in the real and complex matrix-variate cases are derived from the corresponding fractional integrals in Chaps. 3 and 6. Here, the operators introduced can work only on certain types of functions of real and complex matrix argument and hence not universal. This area is open to come up with universal differential operators operating on real-valued functions of real and complex matrix argument.

Montreal, Canada A. M. Mathai
Vienna, Austria H. J. Haubold
20 February 2018

Contents

Acronyms

$\mathrm{d}X, \mathrm{d}\tilde{X}$	: wedge product of differential/Sect. 2.2
$X > O, \tilde{X} > O$	: positive definite matrices; real case, complex case/Sect. 2.2
$\int_A^B f(X)\mathrm{d}X$	: integral over matrices/Sect. 2.2
$\det(X), \lvert X\rvert$	: determinant of X, real case/Sect. 2.2
$\lvert\det(\tilde{X})\rvert$	: absolute value of the determinant of $\tilde{X}$/Sect. 2.2
$D_{1,(a,x)}^{-\alpha} f$	: Riemann-Liouville fractional integral of the first kind/Sect. 2.2
$D_{2,(x,b)}^{-\alpha} f$	: Riemann-Liouvile fractional integral of the second kind/Sect. 2.2
$W_{1,x}^{-\alpha} f$	: Weyl fractional integral of the first kind/Sect. 2.2
$W_{2,x}^{-\alpha} f$	: Weyl fractional integral of the second kind/Sect. 2.2
$K_{1,x,\zeta}^{-\alpha} f$	: Erdélyi-Kober fractional integral of the first kind/Sect. 2.2
$K_{2,x,\zeta}^{-\alpha} f$	: Erdélyi-Kober fractional integral of the second kind/Sect. 2.2
$S_{1,x,\beta,\gamma}^{-\alpha} f$	: Saigo fractional integral of the first kind/Sect. 2.2
$S_{2,x,\beta,\gamma}^{-\alpha} f$	: Saigo fractional integral of the second kind/Sect. 2.2
Replace $-\alpha$ by $+\alpha$	for the corresponding fractional derivatives; replace small x by U_1, U_2 for fractional integrals of the first and second kind in the real matrix-variate cases; replace U_1, U_2 by $\tilde{U}_1, \tilde{U}_2$ for the corresponding fractional integrals in the complex matrix-variate cases
$g_1(u_1), g_{1j}(u_1)$	: density of the ratio or Mellin convolution of a ratio/Sect. 2.3
$E(\cdot)$	: expected value of $(\cdot)$/Sect. 2.3
$g_2(u_2), g_{2j}(u_2)$	: density of a product or Mellin convolution of a product/Sect. 2.9
$A^{\frac{1}{2}}$	: square root of a positive definite matrix A/Sect. 3.1
$\Gamma_p(\alpha)$	: real matrix-variate gamma function/Sect. 3.1
$B_p(\alpha, \beta)$	: real matrix-variate beta function/Sect. 3.1
$g_2(U_2)$	: density of a product or M-convolution of a product, real matrix-variate case/Sect. 3.2

$(a)_K$	: generalized Pochhammer symbol/Sect. 3.5
${}_mF_n$	: hypergeometric series/Sect. 3.5
$C_K(Z)$	: zonal polynomial, real case/Sect. 3.5
$\Gamma_p(\alpha, K)$	: gamma on partitions, real case/Sect. 3.5
$K_{2,u_j,\zeta_j,j=1,\ldots,k}^{-\alpha_j,j=1,\ldots,k} f$	: multivariate second kind Erdélyi-Kober fractional integral/Sect. 4.1
$K_{1,u_j,\zeta_j,j=1,\ldots,k}^{-\alpha_j,j=1,\ldots,k} f$	: multivariate first kind Erdélyi-Kober fractional integral/Sect. 4.4
$\tilde{\Gamma}_p(\alpha)$	: complex matrix-variate gamma function/Sect. 6.1
$\tilde{B}_p(\alpha, \beta)$	: complex matrix-variate beta function/Sect. 6.1
$(a)_K$	: generalized Pochhammr symbol, complex case/Sect. 6.4.3
$\tilde{\Gamma}_p(\alpha, K)$	: gamma on partitions, complex case/Sect. 6.4.3
${}_m\tilde{F}_n$	: hypergeometric series, complex case/Sect. 6.4.3
$\tilde{C}_K(\tilde{Z})$	: zonal polynomial, complex case/Sect. 6.4.3
$\tilde{g}_2(\tilde{U}_2)$	: M-convolution of a product, complex case/Sect. 6.4.3
$\tilde{g}_1(\tilde{U}_1)$	: M-convolution of a ratio, complex case/Sect. 6.5
D_{U_-}, D_{U_+}	: determinant of differential operators, matrix-variate case/Sect. 7.2

Chapter 1
Solar Neutrinos, Diffusion, Entropy, Fractional Calculus

The laws of nature are fundamentally random. This Springer Briefs in Mathematical Physics is an attempt to illustrate elements of a research programme in mathematics and statistics applied to selected problems in physics, particularly the relations between solar neutrinos, diffusion, entropy, and fractional calculus as they appear in neutrino astrophysics since the 1970s. The original research programme was published in three monographs [18–20]. An update of this research programme and selected results achieved since the 1970s is contained in Mathai and Haubold [21] and Mathai, Saxena, and Haubold [22]. The research programme connects mathematics and statistics to theoretical physics with the following historical background in mind.

History has seen a great relation between mathematics and statistics and their impact on physics: Mathematical structures entered the development of theoretical physics or, vice versa, problems aising in physics influenced strongly developments in mathematics and statistics. Famous nineteenth-century and twentieth-century examples are Boltzmann's statistical mechanics and the mathematical concept of entropy, the role of Riemannian geometry in general relativity, and the influence of quantum mechanics in the development of functional analysis. Einstein finalized general relativity in 1915 and quantum field theory has been an open problem since its foundation in 1927 by Dirac. Today there are three fundamental theories in twenty-first century physics: statistical mechanics, general relativity, and quantum field theory. These theories describe the same natural world on very different scales. General relativity describes gravitation on an astronomical scale, quantum field theory describes the interaction of elementary particles through electromagnetic, strong, and weak forces, and statistical mechanics starts from appropriate microscopic laws (classical, relativistic, quantum) and by adequately using probability theory, to ultimately arrive to the thermodynamical relations and laws extended deeper into quantum field theory. The unification of such theories is pursued by mathematicians and physicists so far with some success. Einstein invented general reativity to resolve an inconsistency between special relativity and Newtonian gravity. Quantum field theory was invented to reconcile Maxwell's

A. M. Mathai, H. J. Haubold, *Erdélyi–Kober Fractional Calculus*, SpringerBriefs in Mathematical Physics 31, https://doi.org/10.1007/978-981-13-1159-8_1

electromagnetism and special relativity with nonrelativistic quantum mechanics. Einstein's thought experiments guided the discovery of general relativity based on the mathematics of Riemannian geometry. For quantum field theory experimental results played the important role with no a priori mathematical model available. Boltzmann-Gibbs entropy works perfectly but only within certain limits and if the physical system is out of equilibrium or its component states depend strongly on one another a generalized entropy should be used. Witten [34] summarized this situation by saying that

> Experiment is not likely to provide detailed guidance about reconciliation of general relativity with quantum field theory. One might, therefore, believe that the only hope is to emulate the history of general relativity, inventing by sheer thought a new mathematical framework which will generalize Riemannian geometry and will be capable of encompassing quantum field theory. Many ambitious theoretical physicists have aspired to do such a thing, but little has come of such efforts.

Boltzmann's derivation of the second law of thermodynamics was based on mechanics arguments. In his paper of 1872, Boltzmann considered the dynamics of binary collisions and stated that "One has therefore rigorously proved that, whatever the distribution of the kinetic energy at the initial time might have been, it will, after a very long time, always necessarily approach that found by Maxwell" [1]. Boltzmann's Stosszahlansatz, i.e. the assumption of molecular chaos used in his equation, was a statistical assumption which had no dynamical basis. His equally famous relation between entropy and probability, $S \sim log W$, in his paper "On the relation between the second law of the mechanical theory of heat and probability theory with respect to the laws of thermal equilibrium" [2] was not based on dynamics either. At that time Boltzmanns Stosszahlansatz was heavily criticized by Loschmidts reversibility paradox [3] and Zermelos recurrence paradox [4–6].

In the remarkable year 1900 for physics, Planck elaborated on the connection between entropy and probability based on the universality of the second law of thermodynamics and the established laws of probability and put in writing the final form of the relation between entropy S and permutability $P \sim W$ in its definitive form $S = klog W$. He called k Boltzmann's constant and came to the conclusion that in every finite region of phase space the thermodynamic probability has a finite magnitude limited by h, representing Planck's constant. At this point Planck introduced his quantum hypothesis [31]. Concerning Planck's hypothesis of light quanta he strictly preserved Maxwell's theory in vacuum and applied the quantum hypothesis only to matter that interacts with radiation [28].

Following the above reasoning of Boltzmann, Planck, and Einstein, the research programme referred to above turned to solar neutrino radiation and utilized the statistical methodology developed by Scafetta [30] by evaluating the scaling exponent of the probability density function, through Boltzmann's entropy, of the diffusion process generated by complex fluctuations in the measurements of the solar neutrino flux in the Super-Kamiokande experiment [9, 14, 29, 35]. This turn was justified by earlier explorations of possible solutions to the so-called solar neutrino problem, established in Davis' Homestake experiment [12, 13, 32]. Neutrinos produced in cycles of thermonuclear reactions (proton-proton chain and CNO cycle) in the Sun

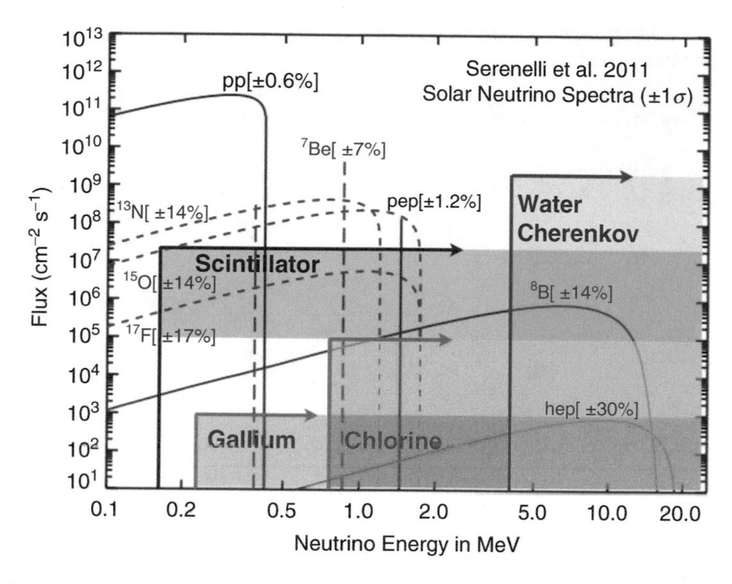

Fig. 1.1 Spectra of neutrinos emitted by fusion reactions in the Sun. Solid lines represent neutrinos from the pp chain and dashed lines are neutrinos from the CNO cycle. Original image taken from Orebi Gann [25] and modified by the author to include the sensitivity of various experimental approaches

can be distinguished by their energy spectra. Figure 1.1 [25] shows the spectrum of neutrinos emitted in each individual reaction: solid lines represent neutrinos from the pp chain, and dashed lines are neutrinos from the CNO cycle. Also shown are the energy regimes in which these neutrinos have been detected [29]. The first experiment to detect neutrinos from the Sun was the Chlorine experiment of Ray Davis et al. at the Homestake mine in South Dakota. These observations were supported by later measurements from gallium-based experiments: GALLEX; SAGE; and GNO. These radio-chemical experiments achieve very low energy thresholds, but perform an integral measurement of all neutrinos above threshold, producing a single integrated flux measurement. Water Cherenkov experiments such as Super-Kamiokande and the Sudbury Neutrino Observatory have higher thresholds but can perform real-time detection thus allowing extraction of both directional and spectral information. This capability allowed Kamiokande I and II to first demonstrate that the observed neutrinos were in fact coming from the Sun. For several decades there was a large discrepancy between the flux of solar neutrinos predicted by the Standard Solar Model and that measured in the above experiments. This became known as the Solar Neutrino Problem. The combination of SNOs measurements with Super-Kamiokandes measurements demonstrated eventually that neutrinos produced in the Sun were oscillating among flavors prior to detection. Scafetta's method does focus on the scaling properties of the Super-Kamiokande time series (see Fig. 1.2) generated by a supposedly unknown complex dynamical phenomenon. By summing the terms of such a time series one gets a trajectory and

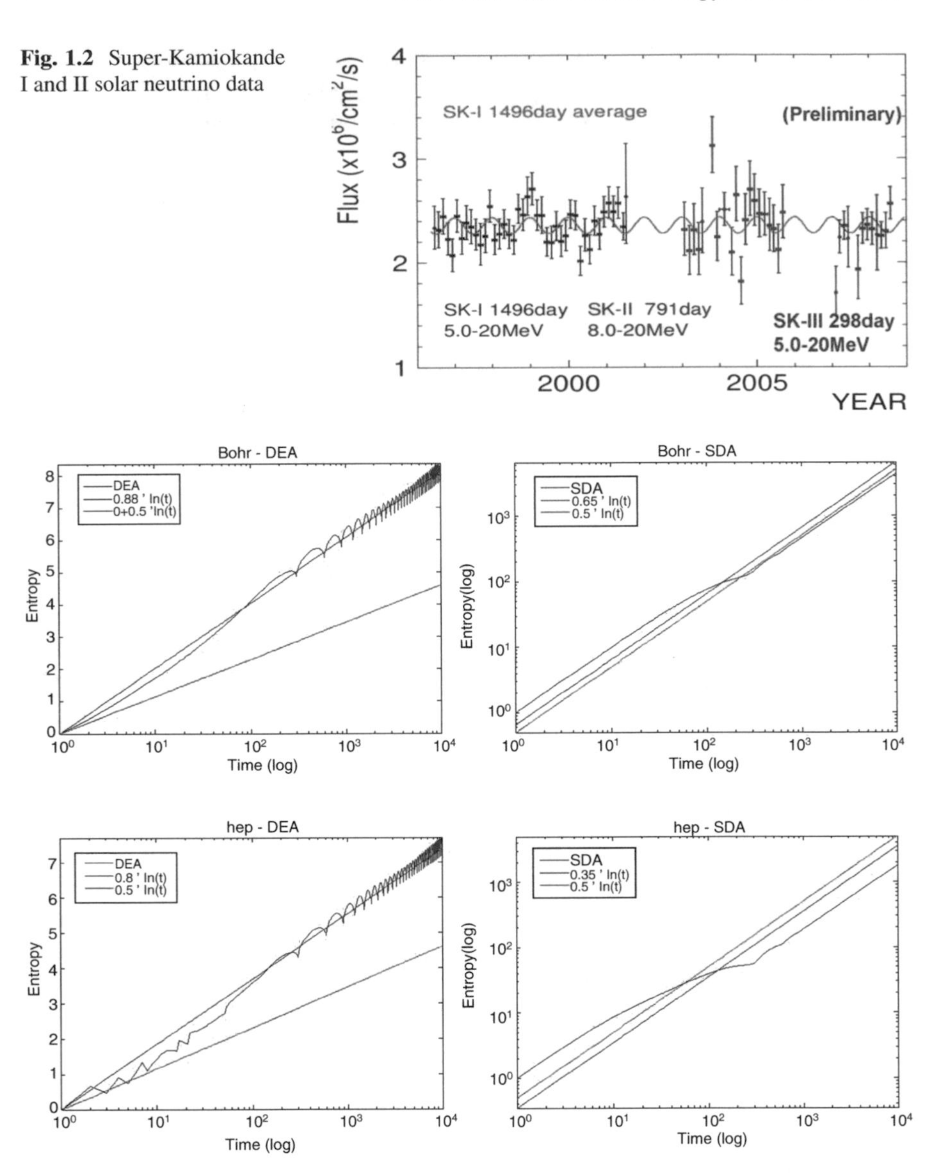

Fig. 1.2 Super-Kamiokande I and II solar neutrino data

Fig. 1.3 Diffusion Entropy Analysis and Standard Deviation Analysis of the Super-Kamiokande I and II solar neutrino data [14]

this trajectory can be used to generate a diffusion process. The method is thus based upon the evaluation of the Boltzmann entropy of the probability density function of a diffusion process. The numerical result of diffusion entropy analysis of the solar neutrino data from Super-Kamiokande is shown in Fig. 1.3.

Todays perception of the quantum mechanics of neutrino flavour oscillations can be analyzed in a variety of ways in physics. There are treatments of this oscillation phenomenon based on plane waves, on wave packets, and on quantum field theory. These treatments have yielded the standard expression for the probability of oscillations. Neutrinos have been detected in three distinct flavours which interact in particular ways with electrons, muons, and tau leptons, respectively. Flavour oscillations occur because the flavour states are distinct from the neutrino mass states. In particular, a given flavour state may be represented as a coherent superposition of different mass states. In the recent MINOS experiment it was discovered that the phenomenon of neutrino oscillations violates the Leggett-Garg inequality, an analogue of Bell's inequality, involving correlations of measurements on neutrino oscillations at different times [11]. The MINOS experiment analysis did show a violation of the classical limits imposed by the Leggett-Garg inequality. This provided evidence for the existence of the quantum effect of entanglement between the mass eigenstates which make up a flavour state. The entropy of entanglement [16, 27] is an entanglement measure for a many-body quantum state and the question arises if the results shown in Fig. 1.3 may find an interpretation in terms of the evolution of an entanglement entropy over time.

In principle, one can perceive the graphical result in Fig. 1.3 of the diffusion entropy analysis (and standard deviation analysis for comparison) of solar neutrino radiation similar to Planck's analysis of black body radiation. What physical meaning this carries remains to be seen. Assuming that the solar neutrino signal is governed by a probability density function (pdf) with scaling given by the asymptotic time evolution of a pdf, obeying the property:

$$p(x,t) = \frac{1}{t^{\delta}} F\left(\frac{x}{t^{\delta}}\right),$$

where δ denotes the scaling exponent of the pdf.

Back to Fig. 1.3, it shows a phenomenon that follows certain scaling laws. This Diffusion Entropy Analysis (DEA) measures the correlated variations in the Super-Kamiokande solar neutrino time series. The analysis is based on the diffusion process generated by the time series and measures the time evolution of the Boltzmann entropy of the probability density function of this diffusion process, possibly a quantum diffusion phenomenon. Similar to Brownian motion trajectories, the value of a time series is interpreted as the steps of a diffusion process. The trajectories of this process are defined by the cumulative sum of these steps and obtain a different trajectory for each value of the time series over the full period of time of measurements. Subsequently the probability density function $p(x, t)$ is evaluated that describes the probability that a given trajectory has a displacement of x after t steps. For every particular t the temporal Boltzmann entropy of the probability density function $p(x, t)$ at time t is evaluated by $S(t) = \delta \log t$, where δ is the diffusion exponent. For a random uncorrelated diffusion process with finite variance, the $p(x, t)$ will converge according to the central limit theorem to a Gaussian pdf which exhibits $\delta = 1/2$. Figure 1.3 shows clearly that all δ's are different from the value $\delta = 1/2$. These diffusion exponents are non-Gaussian and exhibit diffusive fluctuations that cannot be modeled by random Gaussian diffusion processes.

To evaluate the Boltzmann entropy of the diffusion process at time t, [30] defined $S(t)$ as:

$$S(t) = -\int_{-\infty}^{+\infty} dx\ p(x,t) \ln\ p(x,t)$$

and with the previous $p(x,t)$, one has:

$$S(t) = A + \delta \ln(t), \quad A = -\int_{-\infty}^{+\infty} dy F(y) \ln F(y)$$

The scaling exponent, δ, is the slope of the entropy against the logarithmic time scale. The slope is visible in Fig. 1.3 for the Super-Kamiokande data I and II measured for the solar neutrino fluxes generated in ^{8}B and *hep* nuclear reactions in the gravitationally stabilized solar fusion reactor. The Hurst exponents of the Standadrd Deviation Analysis (SDA) of the same time series are $H = 0.66$ and $H = 0.36$ for ^{8}B and *hep*, respectively, shown in Fig. 1.3. The pdf scaling exponents for DEA are $\delta = 0.88$ and $\delta = 0.80$ for ^{8}B and *hep*, respectively. The values for both SDA and DEA indicate a deviation from Gaussian behavior, which would require that $H = \delta = 1/2$.

In 1911 at the first Solvay Conference, Einstein literally put it as an requirement that one needs a fundamental theory of dynamics to make sense of Boltzmann's connection between entropy and probability, even in the case of Planck's use of Boltzmann's formula in the process of discovery of the quantum of action. Einstein's immediate reaction to Planck's extensive report at the Solvay Congress was Eucken [10]:

> What I find strange about the way Mr. Planck applies Boltzmann's equation is that he introduces a state probability W without giving this quantity a physical definition. If one proceeds in such a way, then, to begin with, Boltzmann's equation does not have a physical meaning. The circumstance that W is equated to the number of complexions belonging to a state does not change anything here; for there is no indication of what is supposed to be meant by the statement that two complexions are equally probable. Even if it were possible to define the complexions in such a manner that the S obtained from Boltzmann's equation agrees with experience, it seems to me that with this conception of Boltzmann's principle it is not possible to draw any conclusions about the admissibility of any fundamental theory whatsoever on the basis of the empirically known thermodynamic properties of a system.

Recently, Brush and Segal [8] commented on the above Boltzmann-Planck-Einstein dispute from a historical point of view on how the interaction of theory and experiment in physics with available applicable mathematics and statistics lead to established theories and subsequently to predictions and explanations of natural phenomena. He perceives Planck's derivation of an equation for black-body radiation that this equation, when mated with Boltzmann's formula for entropy, implied that radiation is composed of particles. Planck, as a strong supporter of the wave theory of electromagnetic radiation, could not believe what the mathematics was telling him. Similarly, Kuhn [15] pointed out that Planck did not propose a physical quantum theory but he used quantization only as a convenient method of approximation.

The Boltzmann-Gibbs statistical mechanics exhibits highly relevant connections at the microscopic, mesoscopic, and macroscopic physical levels as well as with the theory of probabilities (Central Limit Theorem). In general, the effects of the Central Limit Theorem with its Gaussian attractors (in the space of the distributions of probabilities) dominate. However, when basic assumptions (molecuar chaos hypothesis, ergodicity) for the applicability of the Boltzmann-Gibbs theory are violated, the concept of entropy needs to be extended. Such extensions were put forward by Mathai's additive [18] and Tsallis non-additive [33] generalizations of Boltzmann-Gibbs entropy.

One of the well known random walk models is the Continuous Time Random Walk (CTRW) introduced by Montroll and Weiss [24]. It describes a large class of random walks, both normal and anomalous, and can be described as follows. Suppose a particle performs a random walk in such a way that the individual jump x in space is governed by a probability density function and that all jumps are independent and identically distributed. The characteristic function of the position of the particle relative to the origin after n jumps is $f^n(k)$, where $f^*(k)$ is the Fourier transform of $f(x)$. Unlike discrete time random walks, the CTRW describes a situation where the waiting time t between jumps is not a constant. Rather, the waiting time is governed by the PDF $\psi(t)$ and all waiting times are mutually independent and identically distributed. Thus, number of jumps n is a random variable. Let $p(x, t)$ be the Green function of the CTRW, the Montroll–Weiss equation yields this function in Fourier–Laplace (k, u) space:

$$p(k, u) = \frac{1 - \psi(u)}{u} \frac{1}{1 - f^*(k)\psi(u)}.$$

All along the above we used the convention that the arguments in the parenthesis define the space we are working in, thus $\psi(u)$ is the Laplace transform of $\psi(t)$. Properties of $p(x, t)$ based on the Fourier–Laplace inversion of the previous equation are well investigated, see Mainardi et al. [17]. In particular, it is well known that the asymptotic behavior of $p(x, t)$ depends on the long time behavior of $\psi(t)$. An important assumption made in the derivation of the previous equation is that the random walk begun at time $t = 0$. More precisely, it is assumed that the pdf of the first waiting time, i.e., the time elapsing between start of the process at $t = 0$ and the first jump event is $\psi(t)$. Thus the Montroll-Weiss CTRW approach describes a particular choice of initial conditions, called non-equilibrium initial conditions.

The following diffusion models (Fig. 1.4) utilize fractional-order spatial and fractional-order temporal derivatives [23]

$${}_0D_t^{\beta}\, p(x, t) = \eta\, {}_xD_{\theta}^{\alpha}\, p(x, t),$$

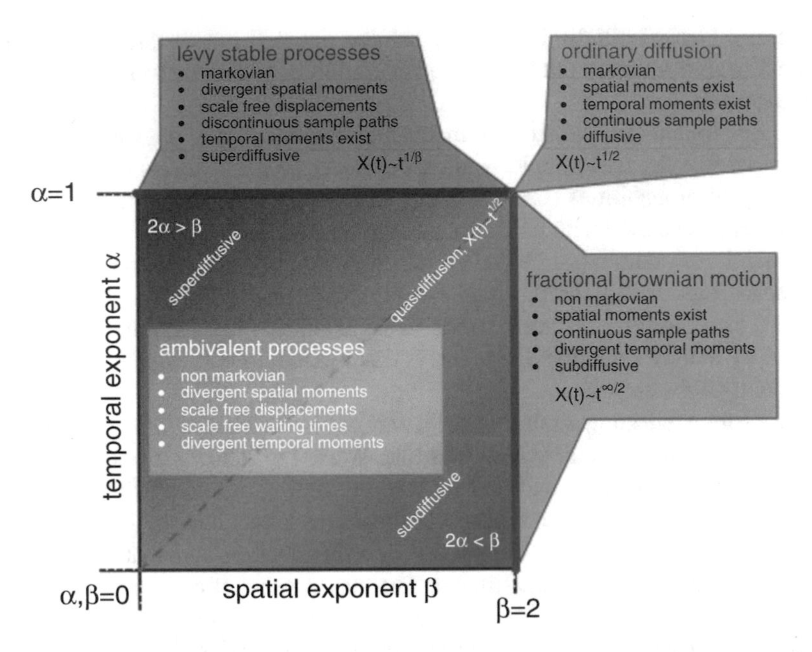

Fig. 1.4 The asymptotic universality classes of continuous time random walks as a function of universality exponents $0 < \alpha < 1$ and $0 < \beta < 2$ [7]

with the initial conditions ${}_0D_t^{\beta-1}p(x,0) = \sigma(x), 0 \leq \beta \leq 1, \lim_{x\to\pm\infty} p(x,t) = 0$, where η is a diffusion constant; $\eta, t > 0, x \in R; \alpha, \theta, \beta$ are real parameters with the constraints $0 < \alpha \leq 2$, $|\theta| \leq min(\alpha, 2-\alpha)$, and $\delta(x)$ is the Dirac-delta function. Then for the fundamental solution of the previous fractional differential equation with initial conditions, there holds the formula

$$p(x,t) = \frac{t^{\beta-1}}{\alpha|x|} H_{3,3}^{2,1}\left[\frac{|x|}{(\eta t^{\beta})^{1/\alpha}}\left|\begin{matrix}(1,1/\alpha),(\beta,\beta/\alpha),(1,\rho)\\(1,1/\alpha),(1,1),(1,\rho)\end{matrix}\right.\right], \alpha > 0$$

where $\rho = \frac{\alpha-\theta}{2\alpha}$, in terms of Fox's H-function. The following special cases of the previous fractional differential equation are of special interest for fractional diffusion models:

(i) For $\alpha = \beta$, the corresponding solution of the fractional differential equation, denoted by p_{α}^{θ}, can be expressed in terms of the H-function and can be defined for $x > 0$:

Non-diffusion: $0 < \alpha = \beta < 2; \theta \leq \min\{\alpha, 2-\alpha\}$,

$$p_{\alpha}^{\theta}(x) = \frac{t^{\alpha-1}}{\alpha|x|} H_{3,3}^{2,1}\left[\frac{|x|}{t\eta^{1/\alpha}}\left|\begin{matrix}(1,1/\alpha),(\alpha,1),(1,\rho)\\(1,1/\alpha),(1,1),(1,\rho)\end{matrix}\right.\right], \quad \rho = \frac{\alpha-\theta}{2\alpha}.$$

(ii) When $\beta = 1, 0 < \alpha \leq 2; \theta \leq \min\{\alpha, 2-\alpha\}$, then the previous fractional differential equation reduces to the space-fractional diffusion equation, which is the fundamental solution of the following space-time fractional diffusion model:

$$\frac{\partial p(x,t)}{\partial t} = \eta\, {}_x D_\theta^\alpha p(x,t), \eta > 0, x \in R,$$

with the initial conditions $p(x, t = 0) = \sigma(x)$, $\lim_{x \to \pm\infty} p(x,t) = 0$, where η is a diffusion constant and $\sigma(x)$ is the Dirac-delta function. Hence for the solution of the previous fractional differential equation there holds the formula

$$p_\alpha^\theta(x) = \frac{1}{\alpha(\eta t)^{1/\alpha}} H_{2,2}^{1,1}\left[\frac{(\eta t)^{1/\alpha}}{|x|}\middle|_{(\frac{1}{\alpha},\frac{1}{\alpha}),(\rho,\rho)}^{(1,1),(\rho,\rho)}\right], \; 0 < \alpha < 1, |\theta| \leq \alpha,$$

where $\rho = \frac{\alpha-\theta}{2\alpha}$. The density represented by the above expression is known as α-stable Lévy density. Another form of this density is given by

$$p_\alpha^\theta(x) = \frac{1}{\alpha(\eta t)^{1/\alpha}} H_{2,2}^{1,1}\left[\frac{|x|}{(\eta t)^{1/\alpha}}\middle|_{(0,1),(1-\rho,\rho)}^{(1-\frac{1}{\alpha},\frac{1}{\alpha}),(1-\rho,\rho)}\right], 1 < \alpha < 2, |\theta| \leq 2-\alpha.$$

(iii) If one takes $\alpha = 2, 0 < \beta < 2; \theta = 0$, then one obtains the time-fractional diffusion, which is governed by the following time-fractional diffusion model:

$$\frac{\partial^\beta p(x,t)}{\partial t^\beta} = \eta \frac{\partial^2}{\partial x^2} p(x,t), \eta > 0, x \in R, \; 0 < \beta \leq 2,$$

with the initial conditions ${}_0D_t^{\beta-1} p(x,0) = \sigma(x), {}_0 D_t^{\beta-2} p(x,0) = 0$, for $x \in r$, $\lim_{x\to\pm\infty} p(x,t) = 0$, where η is a diffusion constant and $\sigma(x)$ is the Dirac-delta function, whose fundamental solution is given by the equation

$$p(x,t) = \frac{t^{\beta-1}}{2|x|} H_{1,1}^{1,0}\left[\frac{|x|}{(\eta t^\beta)^{1/2}}\middle|_{(1,1)}^{(\beta,\beta/2)}\right].$$

(iv) If one sets $\alpha = 2, \beta = 1$ and $\theta \to 0$, then for the fundamental solution of the standard diffusion equation

$$\frac{\partial}{\partial t} p(x,t) = \eta \frac{\partial^2}{\partial x^2} p(x,t),$$

with initial condition $p(x, t = 0) = \sigma(x)$, $\lim_{x\to\pm\infty} p(x,t) = 0$, there holds the formula

$$p(x,t) = \frac{1}{2|x|} H_{1,1}^{1,0}\left[\frac{|x|}{\eta^{1/2} t^{1/2}}\Big|_{(1,1)}^{(1,1/2)}\right] = (4\pi\eta t)^{-1/2} \exp[-\frac{|x|^2}{4\eta t}],$$

which is the classical Gaussian density.

In a different way the previous fractional differential equation for $p(x,t)$ can also be written [26]

$$\frac{\partial p(x,t)}{\partial t} = \frac{2H}{\beta} t^{2H-1} \mathscr{D}_{2H/\beta}^{\beta-1,1-\beta} \frac{\partial^2 p(x,t)}{\partial x^2},$$

where $\mathscr{D}_{\eta}^{\xi,\mu}$ is the Erdélyi–Kober fractional derivative with respect to t and then the process was also referred to as *Erdélyi–Kober fractional diffusion* . Special cases of the previous equation are: the classical diffusion ($\beta = 2H = 1$), the fractional Brownian motion master equation ($\beta = 1$), and the time-fractional diffusion equation ($\beta = 2H$). A similar approach can be developed in the framework of the space-time fractional diffusion equation, which includes all its special cases. Propagation of neutrino radiation may put forward a new class of phenomena that nonequiilibrium quantum systems may exhibit as shown in Fig. 1.3. This could be an Erdélyi-Kober fractional diffusion operator, a mathematical operator that describes the evolution of the probability density function of the quantum system, and the partition function which describes the statitiscal properties of the system in thermal nonequilibrium with the environment. This will be worked out in future research.

References

1. L. Boltzmann, Weitere Studien ueber das Waermegleichgewicht unter Gasmolekuelen, Wiener Berichte **66**, 275–370 (1872), Wissenschaftliche Abhandlungen, Band I, 316–402; English translation: Further studies on the thermal equilibrium of gas molecules, in *Kinetic Theory 2*, ed. by S.G. Brush (Oxford, Pergamon, 1966), pp. 88–174
2. L. Boltzmann, Ueber die Beziehung zwischen dem zweiten Hauptsatz der mechanischen Waermetheorie und der Wahrscheinlichkeitsrechnung respektive den Saetzen ueber das Waermegleichgewicht, Wiener Berichte **76**, 373–435 (1877); Wissenschaftliche Abhandlungen, Band II, 164–223
3. L. Boltzmann, Bemerkungen ueber einige Probleme der mechanischen Waermetheorie, Wiener Berichte **75**, 62–100 (1877); Wissenschaftliche Abhandlungen, Band II, 112–150
4. L. Boltzmann, Entgegnung auf die waermetheoretischen Betrachtungen des Hrn. E. Zermelo, Wiedener Annalen **57**, 773–784 (1896); Wissenschaftliche Abhandlungen, Band III, 567–578
5. L. Boltzmann, Zu Hrn. Zermelos Abhandlung ueber die mechanische Erklaerung irreversibler Vorgaenge, Wiedener Annalen **60**, 392–398 (1897); Wissenschaftliche Abhandlungen, Band III, 579–586
6. L. Boltzmann, Ueber einen mechanischen Satz Poincares, Wiener Berichte **106**, 12–20 (1897); Wissenschaftliche Abhandlungen, Band III, 587–595
7. D. Brockmann, L. Hufnagel, T. Geisel, The scaling laws of human travel. Nature **439**, 462–465 (2006). https://doi.org/10.1038/nature04292

8. S.G. Brush, A. Segal, *Making 20th Century Science: How Theories Became Knowledge* (Oxford University Press, Oxford, 2015)
9. J.P. Cravens et al., Solar neutrino measurements in Super-Kamiokande-II. Phys. Rev. D **78**, 032002, (2008)
10. A. Eucken, Die Theorie der Strahlung und der Quanten, Verhandlungen auf einer von E. Solvay einberufenen Zusammenkunft (30. Oktober bis 3. November 1911). Mit einem Anhang ueber die Entwicklung der Quantentheorie vom Herbst 1911 bis zum Sommer 1913, Druck und Verlag von Wilhelm Knapp, Halle 1914, p. 95
11. J.A. Formaggio, D.I. Kaiser, M.M. Murskyj, T.E. Weiss, Violation of the Leggett-Garg inequality in neutrino oscillations. Phys. Rev. Lett. **117**, 050402 (2017)
12. H.J. Haubold, R.W. John, On the evaluation of an integral connected with the thermonuclear reaction rate in closed form. Astronomische Nachrichten **299**, 225–232 (1978)
13. H.J. Haubold, E. Gerth, The search for possible time variations in Davis' measurements of the argon production rate in the solar neutrino experiment. Astronomische Nachrichten **306**, 203–211 (1985)
14. H.J. Haubold, A.M. Mathai, R.K. Saxena, Analysis of solar neutrino data from Super-Kamiokande I and II. Entropy **16**, 1414–1425 (2014)
15. T.S. Kuhn, *Black-Body Theory and the Quantum Discontinuity 1894–1912* (Clarendon Press, Oxford, 1978)
16. Z.-W. Liu, S. Lloyd, E.Y. Zhu, H. Zhu, Generalized entanglement entropies of quantum design. arXiv: 1709.04313 [quant-ph] (2017)
17. F. Mainardi, Y. Luchko, G. Pagnini, The fundamental solution of the space-time fractional diffusion equation. Fract. Calc. Appl. Analysis **4**, 153–192 (2001)
18. A.M. Mathai, P.N. Rathie, *Basic Concepts in Information Theory and Statistics: Axiomatic Foundations and Applications* (Wiley, New York, 1975)
19. A.M. Mathai, G. Pederzoli, *Characterizations of the Normal Probability Law* (Wiley, New York, 1977)
20. A.M. Mathai, R.K. Saxena, *The H-funtion with Applications in Statistics and Other Disciplines* (Wiley, New York, 1978)
21. A.M. Mathai, H.J. Haubold, *Modern Problems in Nuclear and Neutrino Astrophysics* (Akademie-Verlag, Berlin, 1988)
22. A.M. Mathai, R.K. Saxena, H.J. Haubold, *The H-Function: Theory and Applications* (Springer, New York, 2010)
23. S. Naik, H.J. Haubold, On the q-Laplace transform and related special functions. Axioms **5**(3), 24 (2016). https://doi.org/10.3390/axioms5030024
24. I. Oppenheim, K.E. Shuler, G.H. Weiss, *Stochastic Processes in Chemical Physics: The Master Equation* (MIT Press, Cambridge, 1977)
25. G.D. Orebi Gann, Everything under the Sun: a review of solar neutrinos, in *AIP Conference Proceedings 1666*, 2015, p. 090003. https://doi.org/10.1063/1.4915568
26. G. Pagnini, Erdelyi-Kober fractional diffusion. Fract. Calc. Appl. Analysis **15**, 117–127 (2012)
27. R. Penrose, *Fashion, Faith, and Fantasy in the New Physics of the Universe* (Princeton University Press, Princeton and Oxford, 2016)
28. M. Planck, July 1907 Letter from Planck to Einstein, in *The Collected Papers of Albert Einstein, Volume 5: The Swiss Years, Correspondence 1902–1914*, ed. by M.J. Klein, A.J. Kox, R. Schulmann (Princeton University Press, Princeton, 1995), Document 47
29. K. Sakurai, *Solar Neutrino Problems: How They Were Solved* (TERRAPUB, Tokyo, 2014)
30. N. Scafetta, *Fractal and Diffusion Entropy Analysis of Time Series: Theory, Concepts, Applications and Computer Codes for Studying Fractal Noises and Levy Walk Signals* (VDM Verlag Dr. Mueller, Saarbruecken (Germany), 2010)
31. H.-G. Schoepf, Von Kirchhoff bis Planck, Theorie der Waermestrahlung in historisch-kritischer Darstellung (Akademie-Verlag, Berlin, 1978), pp. 105–127

32. H.-J. Treder, Gravitation und weitreichende schwache Wechselwirkungen bei Neutrino-Feldern (Gedanken zu einer Theorie der solaren Neutrinos). Astronomische Nachrichten **295**, 169–184 (1974)
33. C. Tsallis, Possible generalization of Boltzmann-Gibbs statistics. J. Stat. Phys. **52**, 479–487 (1988)
34. E. Witten, Physics and Geometry, in *Proceedings of the International Congress of Mathematicians*, Berkeley, 1986 (American Mathematical Society, Providence, 1987), pp. 267–303
35. J. Yoo et al., Search for periodic modulations of the solar neutrino flux in Super-Kamiokande-I. Phys. Rev. D **68**, 092002 (2003)

Chapter 2
Erdélyi-Kober Fractional Integrals in the Real Scalar Variable Case

2.1 Introduction

This monograph will examine a new definition for fractional integrals in terms of the distributions of products and ratios of statistically independently distributed positive scalar random variables or in terms of Mellin convolutions of products and ratios in the case of real scalar variables. The idea will be generalized to cover real multivariate cases as well as to real matrix-variate cases. In the matrix-variate case, M-convolutions of products and ratios will be used to extend the ideas. Then we will give a definition for the case of real-valued scalar functions of several real matrices. Then we examine fractional calculus in the complex domain. Here $p \times p$ Hermitian positive definite matrices and real-valued scalar functions of these matrices are examined to define and evaluate fractional integrals. It is shown that one can define all types of fractional integrals and fractional derivatives through Erdélyi-Kober fractional integral operators and statistical distribution theory. Then differential operators will be defined by using the following argument. If $D^{-\alpha}$ denotes fractional integral of order α then D^{α} will be called fractional derivative of order α. For $n = 1, 2, \ldots,$ D^n denotes the integer order derivatives. For $\Re(n - \alpha) > 0$ we can define $D^{\alpha} = D^n D^{-(n-\alpha)}$ or $D^{\alpha} = D^{-(n-\alpha)} D^n$. The first will be fractional derivative of order α in the Riemann-Liouville sense and the latter in the Caputo sense. In the present chapter we concentrate on fractional integrals in the real scalar cases.

We start with the examination of statistical densities of products and ratios involving statistically independently distributed real scalar random variables because the theory to be developed is intertwined with statistical distributions, Mellin convolutions and Erdélyi-Kober fractional integrals. Let x_1 and x_2 be statistically independently distributed real scalar positive random variables and let $u_1 = \frac{x_2}{x_1}$ and $u_2 = x_1 x_2$. Consider the densities of u_1 and u_2. We will show that Erdélyi-Kober fractional integral operator of the second kind or right-sided,

A. M. Mathai, H. J. Haubold, *Erdélyi–Kober Fractional Calculus*, SpringerBriefs in Mathematical Physics 31, https://doi.org/10.1007/978-981-13-1159-8_2

operating on an arbitrary density, is available as the density of u_2 and Erdélyi-Kober fractional integral operator of the first kind or left-sided, operating on an arbitrary density, is available as the density of u_1 when x_1 has a type-1 beta density and x_2 has an arbitrary density. Arbitrary density here means any chosen density, not specified beforehand or any real-valued scalar function $f(x)$ of the real scalar variable x such that $f(x) \geq 0$ for all x and $\int_x f(x)\mathrm{d}x = 1$. We also give interpretations for Erdélyi-Kober fractional integrals of the first kind as Mellin convolutions of ratios and Erdédyi-Kober fractional integrals of the second kind as Mellin convolutions for products. Then various types of generalizations will be given thereby obtaining a large collection of operators and fractional integrals which can all be called generalized Erdélyi-Kober fractional integral operators and fractional integrals. One generalization considered is through the pathway idea [14, 16] where one can move from one family of fractional integrals to another family and yet another family and in the limiting case end up with an exponential form. Common generalizations in terms of a Gauss' hypergeometric series are also given statistical interpretations and put on a more general structure so that the standard generalizations given by various authors, including Saigo operators, are given statistical interpretations and are derivable as special cases of the general structure considered in this chapter.

The literature on fractional calculus in the real scalar variable case is vast. There are various definitions of fractional integrals and fractional derivatives. An insight into the area of fractional calculus of real scalar variables is available from Gorenflo and Mainardi [2], Hilfer [4], Kilbas and Trujillo [5], Kiryakova [6], Mainardi et al. [13], Metzler et al. [22], Miller and Ross [23], Nishimoto [24], Oldham and Spanier [25], Podlubny [27] and Saigo and Kilbas [29]. An attempt is made by the present authors to combine definitions of fractional integrals with the help of Mellin convolutions of product and ratio and statistical densities of product and ratio of statistically independently distributed positive scalar random variables. It is found that this approach enables one to extend the theory of fractional calculus to the complex domain as well as to real and complex matrix-variate cases.

Fractional differential equations have emerged as a new branch of applied mathematics and have been utilized for modeling purposes, particularly in physics. Fractional differential equations are considered as an alternative model to nonlinear differential equations. For that purpose, several different fractional derivatives and integrals were introduced: Riemann-Liouville, Caputo, Hadamard, Gruenwald-Letnikov, Weyl-Riesz, and Erdélyi-Kober [30, 31]. For special values of parameters, such operators can reduce to well-known differential, integro-differential, or integral operators like the differential operators of hyper-Bessel type, the Riemann-Liouville fractional derivative, the Caputo fractional derivative, and the Erdélyi-Kober fractional derivatives and integrals [7, 11, 12]. Particularly appealing cases in physics are methods of approximating equations with Erdélyi-Kober operators which arise in mathematical and statistical descriptions of anomalous diffusion [1, 26, 28]. Generalized fractional Erdélyi-Kober integrals can be interpreted geometrically [3, 17] for applications in particle physics [19].

2.2 Some Notations

We will use the following standard notations. If $X = (x_{ij})$ is a real $m \times n$ matrix of functionally independent or distinct real variables x_{ij}'s, then $\mathrm{d}X$ will stand for the wedge product of the differentials. That is, $\mathrm{d}X = \wedge_{i=1}^{m} \wedge_{j=1}^{n} \mathrm{d}x_{ij}$. Thus, for example, if U is a row or column vector with real elements $u_1, \ldots, u_m$ then $\mathrm{d}U = \mathrm{d}U' = \mathrm{d}u_1 \wedge \mathrm{d}u_2 \wedge \ldots \wedge \mathrm{d}u_m$ where a prime denotes the transpose. Matrices in the complex domain will be denoted by a tilde such as $\tilde{X}, \tilde{Y}, \tilde{Z}$ etc. If $\tilde{X} = X + iY$ where X and Y are real $m \times n$ matrices, $i = \sqrt{(-1)}$, then $\mathrm{d}\tilde{X}$ will be defined as $\mathrm{d}\tilde{X} = \mathrm{d}X \wedge \mathrm{d}Y$. For fractional integrals in the complex domain see Mathai (2013) [17]. In the real matrix-variate case, all the matrices arising are $p \times p$ real positive definite unless stated otherwise. In the real matrix-variate case, the notations $X > O, X \geq O, X < O, X \leq O$ will be used to denote positive definite, positive semi-definite, negative definite, negative semi-definite matrices respectively. Matrices which do not belong to the above four categories are called indefinite matrices. Similar notations will be used for Hermitian positive definite matrices. $X^{\frac{1}{2}}$ will denote the real positive definite square root of the real positive definite matrix X. Then $U_2 = X_2^{\frac{1}{2}} X_1 X_2^{\frac{1}{2}}$ is called a symmetric product of the matrices X_1 and X_2 and $U_1 = X_2^{\frac{1}{2}} X_1^{-1} X_2^{\frac{1}{2}}$ is called a symmetric ratio of X_2 over X_1. $\int_{X>O} f(X)\mathrm{d}X$ will denote the integral over the real-valued scalar function of the real matrix argument X, over all positive definite matrices X. $O < A < X < B$ will imply that $A > O, B > O, X - A > O, B - X > O$, all real positive definite. $\det(X)$ will denote the determinant of X, $\mathrm{tr}(X)$ the trace of X, $|\det(\tilde{X})|$ the absolute value of the determinant of $\tilde{X}$ respectively. That is, for a matrix in the complex domain, if $\det(\tilde{X}) = a + ib, i = \sqrt{(-1)}, a, b$ real scalars then the absolute value $|\det(\tilde{X})| = [(a + ib)(a - ib)]^{\frac{1}{2}} = +[a^2 + b^2]^{\frac{1}{2}}$. Other notations will be described as and when necessary.

2.2.1 *Some of the Fractional Integrals and the Notations in the Literature*

The fractional calculus literature is filled up with all types of notations for various factional integrals and fractional derivatives, introduced by various authors from time to time. Some of the most popular fractional integrals, and the various notations used to denote them, will be listed here for ready reference. The most popular fractional integrals and fractional derivatives are the Riemann-Liouville fractional integrals and the corresponding derivatives.

The left-sided or first kind Riemann-Liouville fractional integral of order α, left limit a

$$ {}_aI_x{}^{\alpha} f = I_{a+}^{\alpha} f = {}_aD_x{}^{-\alpha} f = D_{1,(a,x)}^{-\alpha} f = \frac{1}{\Gamma(\alpha)} \int_a^x (x-t)^{\alpha-1} f(t)\mathrm{d}t \qquad (i) $$

for $\Re(\alpha) > 0$. Here α could be real integer, real fraction, complex number.

The last notation is ours. We will use $-\alpha$ as superscript when it is a fractional integral and α or $+\alpha$ as superscript when it is a fractional derivative of order α. First kind will be indicated by the number 1 as a subscript and the second kind by the number 2. We reserve the letters W for Weyl fractional integral or derivative, K for Erdélyi-Kober case, C for Caputo case, S for Saigo case and D is reserved for Riemann-Liouville case since it is the most popular one. Thus, in our notation (i) will be $D_{1,x}^{-\alpha} f$ if the left limit for the Riemann-Liouville fractional integral of order α is either zero or not specified. If the left limit is to be written then we use $D_{1,(a,x)}^{-\alpha} f$. Here the character in f is unimportant and the parameter x appearing as subscript parameter will be the final variable.

Riemann-Liouville second kind or right-sided fractional integral of order α with the right limit b

$$ {}_xI_b{}^{\alpha} f = I_{b-}^{\alpha} f = {}_xD_b{}^{-\alpha} f = D_{2,(x,b)}^{-\alpha} f = \frac{1}{\Gamma(\alpha)} \int_{t=x}^b (t-x)^{\alpha-1} f(t)\mathrm{d}t, \Re(\alpha) > 0. \qquad (ii) $$

If the right limit is not to be specified then our notation will be $D_{2,x}^{-\alpha} f$.

The first kind or left-sided Weyl fractional integral of order α

$$ {}_{-\infty}I_x{}^{\alpha} f = {}_{-\infty}W_x{}^{-\alpha} f = W_{1,x}^{-\alpha} f = \frac{1}{\Gamma(\alpha)} \int_{-\infty}^x (x-t)^{\alpha-1} f(t)\mathrm{d}t, \Re(\alpha) > 0. \qquad (iii) $$

The last one $W_{1,x}^{-\alpha} f$ is our notation.

The right-sided or second kind Weyl fractional integral of order α

$$ {}_xI_{\infty}{}^{\alpha} f = I_{-}^{\alpha} f = {}_xW_{\infty}{}^{-\alpha} f = W_{2,x}^{-\alpha} f = \frac{1}{\Gamma(\alpha)} \int_x^{\infty} (t-x)^{\alpha-1} f(t)\mathrm{d}t, \Re(\alpha) > 0. \qquad (iv) $$

The last one is our notation. For the Riemann-Liouville and Weyl fractional integral operators denoted above, replace $-\alpha$ by α and vice versa to denote the corresponding fractional derivatives.

The Erdélyi-Kober fractional integral of the first kind or left-sided of order α and parameter ζ

$$I[f(x)] = I[\alpha,\zeta; f(x)] = E^{\alpha,\zeta}_{0,x} f = I^{\zeta,\alpha}_{x} f = (I^{+}_{\zeta,\alpha} f)(x) = K^{-\alpha}_{1,x,\zeta} f$$

$$K^{-\alpha}_{1,x,\zeta} f = \frac{x^{-\alpha-\zeta}}{\Gamma(\alpha)} \int_0^x t^{\zeta}(x-t)^{\alpha-1} f(t)\mathrm{d}t, \Re(\alpha) > 0. \tag{v}$$

The last one in the first line and the left-side on the second line are our notations.

The Erdélyi-Kober fractional integral of the second kind of order α and parameter ζ

$$R[f(x)] = R[\alpha,\zeta; f(x)] = K^{\alpha,\zeta}_{x,\infty} f = (K^{-}_{\zeta,\alpha} f)(x) = (K(\alpha,\zeta; f)(x) = K^{-\alpha}_{2,x,\zeta} f$$

$$K^{-\alpha}_{2,x,\zeta} f = \frac{x^{\zeta}}{\Gamma(\alpha)} \int_x^{\infty} t^{-\zeta-\alpha}(t-x)^{\alpha-1} f(t)\mathrm{d}t, \Re(\alpha) > 0. \tag{vi}$$

The last one in the first line and the left side of the second line are our notations. For Erdélyi-Kober fractional integral operators there are several types of generalizations in single scalar variable case and there are several more notations. The whole thing is a notational jungle. The corresponding fractional derivatives are not in general denoted by replacing $-\alpha$ by $+\alpha$ and vice versa in the above notations. There are additional notations for fractional derivatives.

Saigo left-sided fractional integral of order α and parameters β, γ

$$(I^{\alpha,\beta,\gamma}_{0+} f)(x) = S^{-\alpha}_{1,x,\beta,\gamma} f =$$

$$S^{-\alpha}_{1,x,\beta,\gamma} f = \frac{x^{-\alpha-\beta}}{\Gamma(\alpha)} \int_0^x (x-t)^{\alpha-1} {}_2F_1(\alpha+\beta, -\gamma; \alpha; 1-\frac{t}{x}) f(t)\mathrm{d}t, \Re(\alpha) > 0$$

$$= \frac{\mathrm{d}^n}{\mathrm{d}x^n} (I^{\alpha+n,\beta-n,\gamma-n}_{0+} f)(x), \Re(\alpha) \le 0, [\Re(-\alpha)] + 1 = n. \tag{vii}$$

The last one on the first line and the left side of the second line are our notations. In the above representation $[(\cdot)]$ indicates the integer part of $(\cdot)$.

Saigo right-sided or second kind fractional integral of order α and parameters β, γ

$$(I^{\alpha,\beta,\gamma}_{-} f)(x) = S^{-\alpha}_{2,x,\beta,\gamma} f =$$

$$S^{-\alpha}_{2,x,\beta,\gamma} f = \frac{1}{\Gamma(\alpha)} \int_x^{\infty} (t-x)^{\alpha-1} t^{-\alpha-\beta} {}_2F_1(\alpha+\beta, -\gamma; \alpha; 1-\frac{x}{t}) f(t)\mathrm{d}t, \Re(\alpha) > 0$$

$$= (-1)^n \frac{\mathrm{d}^n}{\mathrm{d}x^n} (I^{\alpha+n,\beta-n,\gamma}_{-} f)(x), \Re(\alpha) \le 0, [\Re(-\alpha)] + 1 = n, \tag{viii}$$

In the Saigo case the corresponding fractional derivatives are denoted by the same symbols as the fractional integrals where I is replaced by D. More notations, more definitions and more generalizations of each of the above, see Mathai et al. [21].

2.3 Fractional Integrals of the First Kind in the Real Scalar Variable Case

Let x_1 and x_2 be statistically independently distributed real positive scalar random variables. Let $u_1 = \frac{x_2}{x_1}$. Let x_1 have a type-1 beta density with parameters (β, α), that is, the density of x_1, denoted by $f_1(x_1)$, is given by

$$f_1(x_1) = \frac{\Gamma(\beta+\alpha)}{\Gamma(\beta)\Gamma(\alpha)} x_1^{\beta-1}(1-x_1)^{\alpha-1},\ 0 < x_1 < 1, \Re(\alpha) > 0, \Re(\beta) > 0. \tag{2.1}$$

Let x_2 have an arbitrary density $f_2(x_2) = f(x_2)$ for some density $f(x_2)$ such that $f(x_2) \geq 0$ for all x_2 and $\int_{x_2} f(x_2)\mathrm{d}x_2 = 1$. Then the density of $u_1 = \frac{x_2}{x_1}$ is available by considering the transformation $u_1 = \frac{x_2}{x_1}$, $v = x_2$. Then $\mathrm{d}x_1 \wedge \mathrm{d}x_2 = -\frac{v}{u_1^2}\mathrm{d}u_1 \wedge \mathrm{d}v$. The joint density of u_1 and v and from there the marginal density of u_1, denoted as $g_1(u_1)$, is available as

$$g_1(u_1) = \int_v f_1(\frac{v}{u_1}) f(v)(-\frac{v}{u_1^2})\mathrm{d}v. \tag{2.2}$$

Limits of u_1 will be from ∞ to v and $0 < v < u_1$. Then the marginal density is the following:

$$g_1(u_1) = \int_{v=0}^{u_1} f_1(\frac{v}{u_1}) f(v)\frac{v}{u_1^2}\mathrm{d}v = \frac{\Gamma(\beta+\alpha)}{\Gamma(\beta)\Gamma(\alpha)} \int_{v=0}^{u_1} (\frac{v}{u_1})^{\beta-1}(1-\frac{v}{u_1})^{\alpha-1}\frac{v}{u_1^2} f(v)\mathrm{d}v.$$

Therefore

$$\frac{\Gamma(\beta)}{\Gamma(\beta+\alpha)} g_1(u_1) = \frac{u_1^{-\beta-\alpha}}{\Gamma(\alpha)} \int_{v=0}^{u_1} (u_1 - v)^{\alpha-1} v^\beta f(v)\mathrm{d}v = K_{1,u_1,\beta}^{-\alpha} f, \tag{2.3}$$

where $K_{1,u_1,\beta}^{-\alpha} f$ denotes the left-sided or first kind Kober fractional integral operator of order α and parameter β, operating on f. The general notation that we will use is the following: For the Erdélyi-Kober operator, letter K is used. For the order, α is used. If it is a fractional integral then $-\alpha$ is written as the superscript to K and if it is derivative then α or $+\alpha$ is written as a superscript. For the first kind or left-sided fractional integral, number 1 is used. The kind 1, variable u_1 and the additional parameter β are written as subscript to K. Thus, Erdélyi-Kober operator of the first kind of order α and parameter β is $K_{1,u_1,\beta}^{-\alpha}$ and this operating on f giving rise to fractional integral of order α, of the first kind and parameter β, as $K_{1,u_1,\beta}^{-\alpha}f$. Note that our variable u_1 is a parameter appearing as a subscript to K and the character in f is unimportant here. We can have the following theorem:

Theorem 2.1 *The Erdélyi-Kober fractional integral operator of the first kind, operating on* f,

$$K_{1,u_1,\beta}^{-\alpha}f = \frac{u_1^{-\beta-\alpha}}{\Gamma(\alpha)}\int_{v=0}^{u_1}(u_1-v)^{\alpha-1}v^{\beta}f(v)\mathrm{d}v \tag{2.4}$$

for $\Re(\beta)>0, \Re(\alpha)>0$, *is available as* $\frac{\Gamma(\beta)}{\Gamma(\beta+\alpha)}g_1(u_1)$ *where* $g_1(u_1)$ *is the density of* $u_1=\frac{x_2}{x_1}$ *where* x_1 *has a type-1 beta density with parameters* (β,α) *and* x_2 *has an arbitrary density, and* x_1 *and* x_2 *are statistically independently distributed.*

When $u_1=\frac{x_2}{x_1}$ where x_1 and x_2 are independently distributed, we have, denoting the expected value of $(\cdot)$ by $E(\cdot)$,

$$\begin{aligned} E(u_1^{s-1}) &= E(x_2^{s-1})E(\frac{1}{x_1})^{s-1} = E(x_2^{s-1})E(x_1^{-s+1}) \\ &= f^*(s)\frac{\Gamma(\beta+\alpha)}{\Gamma(\beta)}\frac{\Gamma(\beta+1-s)}{\Gamma(\alpha+\beta+1-s)} \end{aligned} \tag{2.5}$$

for $\Re(\alpha)>0, \Re(\beta+1-s)>0$ where $f^*(s)$ is the Mellin transform of the arbitrary function $f(x_2)$. Then the Mellin transform of $g_1(u_1)$, with Mellin parameter s, denoted as $g_1^*(s)$ and that of $f_1(x_1)$ as $f_1^*(s)$, we have

$$g_1^*(s) = f^*(s)f_1^*(2-s). \tag{2.6}$$

This means that the Erdélyi-Kober fractional integral operator of the first kind, operating on f, $K_{1,u_1,\beta}^{-\alpha}f$, can be considered as the Mellin convolution of a ratio $u_1=\frac{x_2}{x_1}$. Erdélyi-Kober fractional integral operator of the second kind operating on f can be considered as a Mellin convolution of a product $u_2=x_1x_2$.

Hereafter the notation u_1 will be used for the ratio $u_1=\frac{x_2}{x_1}$ and u_2 for the product $u_2=x_1x_2$. If additional sets of x_1 and x_2 are considered then for the j-th set we will use the notation u_{1j} for the ratio and u_{2j} for the product. If x_1 and x_2 independently distributed positive real scalar random variables then the densities of u_1 and u_2 will be denoted by $g_1(u_1)$ and $g_2(u_2)$ respectively. If the densities of the j-th set of variables are involved then we will use the notation $g_{1j}(u_1)$ for the density of the ratio and $g_{2j}(u_2)$ for the density of the product.

Note from (2.3) that

$$\Gamma(\alpha)u_1^{\beta+\alpha}K_{1,u_1,\beta}^{-\alpha}f = \int_{v=0}^{u_1}v^{\beta}(u_1-v)^{\alpha-1}f(v)\mathrm{d}v. \tag{2.7}$$

This is Euler transform of $f(v)$, see for example Mathai et al. [21]. Note that all transforms where the basic function is a type-1 beta form or either of the form $x^{\alpha}(1-x)^{\beta}$, $0 < x < 1$ or of the form $(x-a)^{\alpha}(b-x)^{\beta}$, $a < x < b$ can be connected to Erdélyi-Kober fractional integral operators.

2.4 A Pathway Generalization of Erdélyi-Kober Fractional Integral Operator of the First Kind

Here the steps are parallel to those of the pathway extension of Erdélyi-Kober fractional integral operator of the second kind. Let the density of x_1 be of the form

$$f_{11}(x_1) = c_1 x_1^{\gamma-1}[1-a(1-q)x_1^{\delta}]^{\frac{\eta}{1-q}}, \tag{2.8}$$

for $1-a(1-q)x^{\delta} > 0, \eta > 0, q < 1, \delta > 0, a > 0$, where

$$c_1 = \frac{\delta[a(1-q)]^{\frac{\gamma}{\delta}}\Gamma(\frac{\eta}{1-q}+1+\frac{\gamma}{\delta})}{\Gamma(\frac{\gamma}{\delta})\Gamma(\frac{\eta}{1-q}+1)}.$$

Then let $u_1 = \frac{x_2}{x_1}$ where x_1 has the pathway density $f_{11}(x_1)$ and x_2 has the density $f_{21}(x_2) = f(x_2)$ where x_1 and x_2 are statistically independently distributed and f is an arbitrary density. Then the density of u_1, denoted by $g_{11}(u_1)$, is the following:

$$\begin{aligned}
g_{11}(u_1) &= \int_v f_{11}(\frac{v}{u_1})f_{21}(v)\frac{v}{u_1^2}\mathrm{d}v \\
&= c_1\int_v (\frac{v}{u_1})^{\gamma-1}[1-a(1-q)(\frac{v}{u_1})^{\delta}]^{\frac{\eta}{1-q}} f_{21}(v)\frac{v}{u_1^2}\mathrm{d}v \\
&= c_1 u_1^{-\gamma-(\frac{\delta\eta}{1-q}+1)}\int_v v^{\gamma}[u_1^{\delta}-a(1-q)v^{\delta}]^{\frac{\eta}{1-q}} f(v)\mathrm{d}v.
\end{aligned} \tag{2.9}$$

Through q one has a collection of operators from (2.9), which can be treated as a generalization of Erdélyi-Kober fractional integral operator of the first kind.

2.5 Some Special Cases

Case (1): For $a = 1, q = 0, \delta = 1, \frac{\eta}{1-q} = \alpha - 1$ we have the following:

$$g_{11}(u_1) = \frac{\Gamma(\gamma+\alpha)}{\Gamma(\gamma)\Gamma(\alpha)}u_1^{-\gamma-\alpha}\int_{v=0}^{u_1} v^{\gamma}(u_1-v)^{\alpha-1}f(v)\mathrm{d}v \tag{2.10}$$

for $\Re(\alpha) > 0, \Re(\gamma) > 0$. Therefore for $\Re(\alpha) > 0, \Re(\gamma) > 0$

$$\frac{\Gamma(\gamma)}{\Gamma(\alpha+\gamma)} g_{11}(u_1) = \frac{u_1^{-\gamma-\alpha}}{\Gamma(\alpha)} \int_{v=0}^{u_1} v^{\gamma}(u_1 - v)^{\alpha-1} f(v)\mathrm{d}v = K_{1,u_1,\gamma}^{-\alpha} f. \tag{2.11}$$

Note that (2.9) gives a generalization of Erdélyi-Kober fractional integral operator of the first kind operating on f. This generalization also gives a path through q. For various values of q one has a collection of functions which can all be considered as generalizations of Erdélyi-Kober fractional integral operators of the first kind operating on f. Also, (2.9) can be looked upon as a generalization of the pathway integral transform introduced in Mathai et al. [20]. Observe that (2.9) can also be looked upon as a generalized pathway transform of $f(x)$. Finally, when $q \to 1_-$ we have the following form:

Case (2):

$$\lim_{q \to 1_-} g_{11}(u_1) = \delta \frac{(a\eta)^{\frac{\gamma}{\delta}}}{\Gamma(\frac{\gamma}{\delta})} u_1^{-\gamma-1} \int_{v=0}^{\infty} v^{\gamma} \mathrm{e}^{-a\eta(\frac{v}{u_1})^{\delta}} f(v)\mathrm{d}v. \tag{2.12}$$

This integral part when $\delta = 1$ is the Laplace transform of $v^{\gamma} f(v)$ with Laplace parameter $\frac{a\eta}{u_1}$.

All the functions described from (2.9), (2.10) and (2.11) can be taken as generalizations of Erdélyi-Kober fractional integral operators of the first kind operating on f. Observe that in the limiting case the fractional nature of the integral is lost.

Case (3): For $a = 1, q = 0, \delta = m, \frac{\eta}{1-q} = \alpha - 1$ we have a special case: $\frac{\Gamma(\gamma)}{\Gamma(\gamma+\alpha)} g_{11}(u_1)$ is (2.6.8) of Mathai and Haubold [18].

2.6 Erdélyi-Kober Fractional Integrals of the First Kind and Hypergeometric Series

Let us append a hypergeometric series to our basic function $x_1^{\beta-1}(1 - x_1)^{\alpha-1}$. Consider the hypergeometric series, for $q \geq p$ or $p = q + 1$ and $|ax| < 1$,

$$\begin{aligned} &{}_pF_q(a_1, \ldots, a_p; \quad b_1, \ldots, b_q; ax_1)x_1^{\beta-1}(1 - x_1)^{\alpha-1} \\ &= \sum_{k=0}^{\infty} \frac{(a_1)_k \ldots (a_p)_k}{(b_1)_k \ldots (b_q)_k} \frac{a^k x_1^k}{k!} x_1^{\beta-1}(1 - x_1)^{\alpha-1} \end{aligned}$$

for $0 < x_1 < 1$ and zero elsewhere. The integral part is the following:

$$\int_0^1 x_1^{\beta+k-1}(1 - x_1)^{\alpha-1}\mathrm{d}x_1 = \frac{\Gamma(\beta+k)\Gamma(\alpha)}{\Gamma(\alpha+\beta+k)} = \frac{\Gamma(\alpha)\Gamma(\beta)}{\Gamma(\alpha+\beta)} \frac{(\beta)_k}{(\alpha+\beta)_k}, \tag{2.13}$$

for $\Re(\alpha) > 0, \Re(\beta) > 0$. Then the integral over the appended hypergeometric series gives $c^{(1)}$ where

$$c^{(1)} = {}_{p+1}F_{q+1}(a_1, \ldots, a_p, \beta; b_1, \ldots, b_q, \alpha + \beta; a)\frac{\Gamma(\alpha)\Gamma(\beta)}{\Gamma(\alpha + \beta)}. \tag{2.14}$$

We can assume all parameters a_j's, b_j's and a to be positive and $\alpha > 0, \beta > 0$ in order to assure positivity. Then the density of $u_1 = \frac{x_2}{x_1}$, denoted by $g_{12}(u_1)$, is given by

$$g_{12}(u_1) = \frac{1}{c^{(1)}}\sum_{k=0}^{\infty}\frac{(a_1)_k \ldots (a_p)_k}{(b_1)_k \ldots (b_q)_k}\frac{a^k}{k!}\int_v (\frac{v}{u_1})^{\beta+k-1}(1 - \frac{v}{u_1})^{\alpha-1}\frac{v}{u_1^2}f(v)\mathrm{d}v.$$

The integral part is the following:

$$u_1^{-\beta-\alpha}\int_{v=0}^{u_1} v^{\beta}(u_1 - v)^{\alpha-1}(\frac{v}{u_1})^k f(v)\mathrm{d}v.$$

$$g_{12}(u_1) = \frac{1}{c^{(1)}}u_1^{-\beta-\alpha}\int_{v=0}^{u_1} v^{\beta}(u_1 - v)^{\alpha-1}{}_pF_q(a_1, \ldots, a_p; b_1, \ldots, b_q; a\frac{v}{u_1})f(v)\mathrm{d}v. \tag{2.15}$$

A particular case of (2.15) in terms of a ${}_2F_1$ is (2.7.1) of Mathai and Haubold [18]. This particular case was given by others earlier as generalization of Erdélyi-Kober fractional integral operator of the first kind operating on f, not as a statistical density or as a Mellin convolution of a ratio. We may replace the argument ax_1 by $a^{\delta_1}x_1^{\delta_2}$ in the hypergeometric function, for $\delta_1 > 0, \delta_2 > 0$, to get more general forms of (2.15).

Instead of a ${}_pF_q$ with argument ax_1 let us append $x_1^{\beta-1}(1 - x_1)^{\alpha-1}$ with a hypergeometric function ${}_pF_q$ with argument $a(1 - x_1)$. Then, proceeding as before we get the normalizing constant as

$$c^{(2)} = {}_{p+1}F_{q+1}(a_1, \ldots, a_p, \alpha; b_1, \ldots, b_q, \alpha + \beta; a)\frac{\Gamma(\beta)\Gamma(\alpha)}{\Gamma(\alpha + \beta)}, \tag{2.16}$$

for $\Re(\alpha) > 0, \Re(\beta) > 0$. Then the density of $u_1 = \frac{x_2}{x_1}$, denoted by $g_{13}(u_1)$, is given by the following:

$$g_{13}(u_1) = \frac{1}{c^{(2)}}\sum_{k=0}^{\infty}\frac{(a_1)_k \ldots (a_p)_k}{(b_1)_k \ldots (b_q)_k}\frac{a^k}{k!}\int_v (\frac{v}{u_1})^{\beta-1}(1 - \frac{v}{u_1})^{\alpha+k-1}(-\frac{v}{u_1^2})f(v)\mathrm{d}v.$$

Then

$$g_{13}(u_1) = \frac{u_1^{-\beta-\alpha}}{c^{(2)}}\int_{v=0}^{u_1} v^{\beta}(u_1 - v)^{\alpha-1}{}_pF_q(a_1, \ldots, a_p; b_1, \ldots, b_q; a(1 - \frac{v}{u_1}))f(v)\mathrm{d}v. \tag{2.17}$$

A particular case of (2.17) in terms of a ${}_2F_1$ is (2.7.3) of Mathai and Haubold [18]. This particular case was given by others earlier as a generalization of Erdélyi-Kober fractional integral of the first kind. The argument in the hypergeometric function could have been $a^{\delta_1}(1-x_1)^{\delta_2}$ for $\delta_1>0, \delta_2>0$ to produce a more general case. Also, one could have taken the argument as $a^{\delta_1}x_1^{\delta_2}(1-x_1)^{\delta_3}$. In all such cases, $g_{13}(u_1)$ as a density makes sense and at the same time keeping the basic structure of Erdélyi-Kober fractional integral of the first kind intact. A particular case of (2.17) is the Saigo operator of the first kind when the ${}_pF_q$ is replaced by a ${}_2F_1$. The main advantage of (2.17) is that it is a direct generalization of Erdélyi-Kober fractional integral of the first kind and at the same time it is a statistical density of a ratio of two statistically independently distributed real positive random variables.

2.7 Mellin Transform of the Generalized Erdélyi-Kober Fractional Integral of the First Kind

Let us consider the Mellin transform in (2.15) and (2.17). The basic integral is the following and will be evaluated through interchange of integrals:

$$\int_0^\infty u_1^{s-1}[u_1^{-\beta-\alpha}\int_{v=0}^{u_1} v^\beta(u_1-v)^{\alpha-1}\frac{v^k}{u_1^k}\mathrm{d}u_1]\mathrm{d}v$$

$$=\int_{v=0}^\infty f(v)v^{\beta+k}[\int_{u_1=v}^\infty u_1^{s-1-\beta-\alpha-k}(u_1-v)^{\alpha-1}\mathrm{d}u_1]\mathrm{d}v$$

$$=\int_{v=0}^\infty f(v)v^{\beta+k}[\int_{y=0}^\infty y^{\alpha-1}(y+v)^{s-1-\beta-\alpha-k}\mathrm{d}y]\mathrm{d}v$$

$$=\int_{v=0}^\infty v^{s-1}f(v)\mathrm{d}v\int_{z=0}^\infty z^{\alpha-1}(1+z)^{-(\alpha+\beta+k+1-s)}\mathrm{d}z$$

$$=f^*(s)\frac{\Gamma(\alpha)\Gamma(\beta+1-s)}{\Gamma(\alpha+\beta+1-s)}\frac{(\beta+1-s)_k}{(\alpha+\beta+1-s)_k},$$

for $\Re(\alpha)>0, \Re(\beta+1-s)>0$. Therefore

$$M\{g_{12}(u_1) \text{ of } (2.15); s\}=\frac{\Gamma(\alpha)}{c^{(2)}}\frac{\Gamma(\beta+1-s)}{\Gamma(\alpha+\beta+1-s)}$$
$$\times {}_{p+1}F_{q+1}(a_1,\ldots,a_p,\beta+1-s;b_1,\ldots,b_q,\alpha+\beta+1-s;a)f^*(s) \tag{2.18}$$

for $\Re(\alpha)>0, \Re(s)<\Re(\beta+1), a>0$. The basic integral to be evaluated in corresponding Mellin transform of $g_{13}(u_1)$ in (2.17) is the following:

$$\int_{u_1=0}^{\infty} u_1^{s-1-\beta-\alpha}[\int_{v=0}^{u_1} v^{\beta}(u_1-v)^{\alpha-1}(1-\frac{v}{u_1})^k f(v)\mathrm{d}u_1]\mathrm{d}v$$
$$=\int_{v=0}^{\infty} f(v)v^{\beta}[\int_{u_1=v}^{\infty} u_1^{s-1-\beta-\alpha}(u_1-v)^{\alpha-1}(1-\frac{v}{u_1})^k\mathrm{d}u_1]\mathrm{d}v$$
$$=\int_{v=0}^{\infty} v^{s-1}f(v)\mathrm{d}v\int_{z=0}^{\infty} z^{\alpha+k-1}(1+z)^{-(\alpha+\beta+1-s+k)}\mathrm{d}z$$
$$=f^*(s)\frac{\Gamma(\alpha)\Gamma(\beta+1-s)}{\Gamma(\alpha+\beta+1-s)}\frac{(\alpha)_k}{(\alpha+\beta+1-s)_k}$$

for $\Re(\alpha)>0, \Re(\beta+1-s)>0$. Therefore

$$M\{g_{13}(u_1)\text{ of (2.17)}; s\}=\frac{\Gamma(\alpha)\Gamma(\beta+1-s)}{\Gamma(\alpha+\beta+1-s)}\times{}_{p+1}F_{q+1}(a_1,\ldots,a_p,\alpha;b_1,\ldots,b_q,\alpha+\beta+1-s;a) \tag{2.19}$$

for $\Re(\alpha)>0, \Re(\beta+1-s)>0, a>0$.

2.8 Riemann-Liouville Operators as Mellin Convolution

Consider the Mellin convolution of a ratio where $f_3(x_1)=\frac{x_1^{-\alpha-1}(1-x_1)^{\alpha-1}}{\Gamma(\alpha)}$ and $f_4(x_2)=x_2^{\alpha}f(x_2)$. Let $u_1=\frac{x_2}{x_1}, v=x_2, x_1=\frac{v}{u_1}$. Then the Mellin convolution of a ratio, denoted by $g_{14}(u_1)$, is the following:

$$g_{14}(u_1)=\int_v f_7(\frac{v}{u_1})f_8(v)\frac{v}{u_1^2}\mathrm{d}v=\frac{1}{\Gamma(\alpha)}\int_{v=0}^{u_1}(u_1-v)^{\alpha-1}f(v)\mathrm{d}v={}_0D_x^{-\alpha}f=D_{1,x}^{-\alpha}f \tag{2.20}$$

is the left-sided Riemann-Liouville fractional integral operator of order α operating on f. Thus, the left-sided Riemann-Liouville fractional integral operator can be considered to be the Mellin convolution of a ratio. Consider the Mellin transform of $g_{14}(u_1)$ of (2.20). That is, evaluating through interchange of integrals,

$$M\{g_{14};s\}=\int_0^{\infty} u_1^{s-1}g_{14}(u_1)\mathrm{d}u_1$$
$$=\int_{u_1=0}^{\infty} u_1^{s-1}[f_3(\frac{v}{u_1})f_4(v)(-\frac{v}{u_1^2})\mathrm{d}v]\mathrm{d}u_1$$
$$=\int_{v=0}^{\infty} f_4(v)[\int_{u_1=v}^{\infty} u_1^{s-1}f_3(\frac{v}{u_1})(-\frac{v}{u_1^2})\mathrm{d}u_1]\mathrm{d}v.$$

Put $x_1 = \frac{v}{u_1}, -\frac{v}{u_1^2}\mathrm{d}u_1 = \mathrm{d}x_1$. Then

$$\int_{u_1=v}^{\infty} u_1^{s-1} f_1(\frac{v}{u_1})(-\frac{v}{u_1^2})\mathrm{d}u_1 = v^{s-1}\int_0^1 \frac{1}{x_1^{s-1}} f_3(x_1)\mathrm{d}x_1 = M\{f_3; 2-s\}v^{s-1}$$

for $f_3(x_1) = 0$ outside the interval [0, 1].

$$\int_{v=0}^{\infty} v^{s-1} f_4(v)\mathrm{d}v = M\{f_4; s\}.$$

Then

$$M\{g_{14}; s\} = M\{f_3; 2-s\}M\{f_4; s\} \tag{2.21}$$

or the right side is of the form

$$\int_{x_1}\int_{x_2} (\frac{x_2}{x_1})^{s-1} f_3(x_1)f_4(x_2)\mathrm{d}x_1 \wedge \mathrm{d}x_2 = \int_{x_1} \frac{1}{x_1^{s-1}} f_3(x_1)\mathrm{d}x_1 \int_{x_2} x_2^{s-1} f_4(x_2)\mathrm{d}x_2 \tag{2.22}$$

or in the form of a Mellin convolution for a ratio.

Theorem 2.2 *The left-sided Riemann-Liouville fractional integral is the Mellin convolution of a ratio $u_1 = \frac{x_2}{x_1}$ when the joint function of x_1 and x_2 is of the form $f_3(x_1)f_4(x_2)$ where*

$$f_3(x_1) = \frac{x_1^{-\alpha-1}}{\Gamma(\alpha)}(1-x_1)^{\alpha-1}, 0 < x_1 < 1$$

and zero elsewhere, and $f_4(x_2) = x_2^{\alpha} f(x_2)$ where $f(x_2)$ is an arbitrary function, such that the Mellin transforms of $f_3(x_1)$ and $f_4(x_2)$ exist. That is,

$$g_{14}(u_1) = \int_v f_3(\frac{v}{u_1})f_4(v)(-\frac{v}{u_1^2})\mathrm{d}v = {}_0D_x^{-\alpha} f = D_{1,(0,x)}^{-\alpha} f = D_{1,x}^{-\alpha} f. \tag{2.23}$$

Note 2.1 Note that in this case $f_3(x_1)$ is not a constant multiple of a statistical density because the exponent of x_1 is $-\alpha - 1$ where $\Re(-\alpha) < 0$. Without loss of generality, $f_4(x_2)$ can be taken as a statistical density. The Mellin transform of ${}_0D_x^{-\alpha} f = D_{1,(0,x)}^{-\alpha} f$ is available in the literature, see for example Mathai and Haubold [18].

$$M\{{}_0D_x^{-\alpha} f; s\} = M\{D_{1,(0,x)}^{-\alpha} f; s\} = \frac{\Gamma(1-\alpha-s)}{\Gamma(1-s)} f^*(\alpha+s), \tag{2.24}$$

for $\Re(s) < 1, \Re(\alpha + s) < 1$ where $f^*(s)$ is the Mellin transform of $f(x)$. Thus, if f is replaced by $x^{-\alpha} f$ then we have

$$M\{({}_0D_x^{-\alpha} x^{-\alpha} f)(x); s\} = \frac{\Gamma(1-\alpha-s)}{\Gamma(1-s)} f^*(s), \Re(s) < 1, \Re(\alpha + s) < 1. \tag{2.25}$$

2.9 Distribution of a Product and Erdélyi-Kober Operators of the Second Kind

Let x_1 and x_2 be real positive scalar random variables which are independently distributed with density functions $f_5(x_1)$ and $f_6(x_2)$ respectively. Then the densities of the product $u_2 = x_1 x_2$ and ratio $u_1 = \frac{x_2}{x_1}$, will be denoted by $g_2(u_2)$ and $g_1(u_1)$ respectively, where

$$g_2(u_2) = \int_v \frac{1}{v} f_5(\frac{u}{v}) f_6(v)\mathrm{d}v = \int_y \frac{1}{y} f_5(y) f_6(\frac{u}{y})\mathrm{d}y. \tag{2.26}$$

Let f_5 be a type-1 beta density and f_6 be an arbitrary density, denoted by $f(x_2)$, arbitrary density in the sense any function $f(x)$ such that $f(x) \geq 0$ for all x and $\int_x f(x)\mathrm{d}x = 1$. Then $g_2(u_2)$ will take an interesting form. Let

$$f_5(x_1) = \frac{\Gamma(\beta+1+\alpha)}{\Gamma(\beta+1)\Gamma(\alpha)} x_1^{\beta} (1-x_1)^{\alpha-1},\ 0 < x_1 < 1, \Re(\alpha) > 0, \Re(\beta) > -1$$

and $f_5(x_1) = 0$ elsewhere. In statistical problems, usually the parameters are real but the results will hold for complex parameters and hence we list the conditions for complex parameters. Then from (2.26) we have

$$\begin{aligned} g_2(u_2) &= \frac{\Gamma(\alpha+\beta+1)}{\Gamma(\beta+1)\Gamma(\alpha)} \int_{t=u_2}^{\infty} \frac{1}{t} (\frac{u_2}{t})^{\beta} (1 - \frac{u_2}{t})^{\alpha-1} f(t)\mathrm{d}t \\ &= \frac{\Gamma(\alpha+\beta+1)}{\Gamma(\beta+1)} \frac{u_2^{\beta}}{\Gamma(\alpha)} \int_{u_2}^{\infty} (t-u_2)^{\alpha-1} t^{-\beta-\alpha} f(t)\mathrm{d}t \\ &= \frac{\Gamma(\alpha+\beta+1)}{\Gamma(\beta+1)} K_{2,u_2,\beta}^{-\alpha} f, \end{aligned}$$

where $K_{2,u_2,\beta}^{-\alpha} f$ denotes the usual Kober fractional integral operator of order α and parameter β and of the second kind operating on f, available in the literature. Hereafter, Erdélyi-Kober operators will be denoted by K, order will be denoted by α; if it a fractional integral then the order will be written as $-\alpha$ and as a superscript and if it is a fractional derivative then the order will be written as superscript α or $+\alpha$; if the fractional integral is of the second kind or right-sided then 2 will be written as a subscript; if it is of the first kind or left-sided then 1 will be written as a subscript; the additional parameter β in Erdélyi-Kober operator and the variable u_2 will be written as subscripts; thus the operator, operating on f, is written as $K_{2,u_2,\beta}^{-\alpha} f$. This means that in terms of a statistical density

$$K_{2,u_2,\beta}^{-\alpha} f = \frac{\Gamma(\beta+1)}{\Gamma(\alpha+\beta+1)} g_2(u_2), \Re(\alpha) > 0, \Re(\beta) > 0, \tag{2.27}$$

where $u_2 = u = x_1 x_2$, the product, with density $g_2(u)$. Then, we have the following result:

Theorem 2.3 *Erdélyi-Kober fractional integral operator of the second kind, operating on a real-valued scalar function f, is a constant multiple of the density of a product of two real scalar statistically independently distributed positive random variables x_1 and x_2 where x_1 has a type-1 beta density with the parameters $(\beta+1, \alpha)$, and x_2 has an arbitrary density $f(x_2)$.*

Then looking at $u_2 = x_1 x_2$ from the point of view of Mellin transforms, we have the following in terms of expected values or statistical expectations, denoted by $E(\cdot)$. Since x_1 and x_2 are independently distributed, we have

$$E(u^{s-1}) = E(x_1^{s-1}) E(x_2^{s-1}). \tag{2.28}$$

This Eq. (2.28), if interpreted in terms of Mellin transform with Mellin parameter s then we have

$$M_{g_2}(s) = M_{f_1}(s) M_{f_2}(s) \tag{2.29}$$

where

$$M_{g_2}(s) = \int_0^\infty u_2^{s-1} g_2(u_2) \mathrm{d}u_2, \ M_{f_j}(s) = \int_0^\infty x_j^{s-1} f_j(x_j) \mathrm{d}x_j, \ j = 1, 2$$

whenever the integrals are convergent. But

$$\begin{aligned} E(x_1^{s-1}) &= \frac{\Gamma(\beta+1+\alpha)}{\Gamma(\beta+1)\Gamma(\alpha)} \int_0^1 x_1^{s-1} x_1^{\beta} (1-x_1)^{\alpha-1} \mathrm{d}x_1 \\ &= \frac{\Gamma(\alpha+\beta+1)}{\Gamma(\beta+1)} \frac{\Gamma(\beta+s)}{\Gamma(\alpha+\beta+s)} \end{aligned}$$

for $\Re(\beta) > 0, \Re(\beta+s) > 0, \Re(\alpha) > 0$ and let $E(x_2^{s-1}) = f^*(s) =$ the Mellin transform of $f(x)$. If the Mellin transform of $g_2(u_2)$, with Mellin parameter s, is denoted by $M\{g_2(u_2); s\} = M_{g_2}(s)$, then

$$M\{\frac{\Gamma(\beta+1)}{\Gamma(\alpha+\beta+1)} g_2(u_2); s\} = \frac{\Gamma(\beta+s)}{\Gamma(\alpha+\beta+s)} f^*(s) = M\{K_{2,u_2,\beta}^{-\alpha} f; s\}. \tag{2.30}$$

This is Mellin convolution of a product. Thus, Erdélyi-Kober fractional integral operator of the second kind operating on f can be considered as a Mellin convolution for a product. The inverse Mellin transform of (2.30) provides explicit expression for Erdélyi-Kober fractional integral operator of the second kind operating on f, namely,

$$K_{2,u_2,\beta}^{-\alpha} f = \frac{1}{2\pi i}\int_{c-i\infty}^{c+i\infty} \frac{\Gamma(\beta+s)}{\Gamma(\alpha+\beta+s)} f^*(s)u^{-s}\mathrm{d}s \tag{2.31}$$

where the form is available through the convolution integral coming from the Mellin convolution of a product, that is, a type-1 beta form with parameters $(\beta+1,\alpha)$ convoluted with the arbitrary function $f(x)$.

If $f_5(x_1)$ is a more general density than a type-1 beta density then we can get some generalizations of Theorem 2.3. We will consider a pathway extension of type-1 beta density first.

2.10 A Pathway Extension of Erdélyi-Kober Operator of the Second Kind

Let $f_{11}(x_1)$ be the pathway density

$$f_{11}(x_1) = c_{11}\, x_1^{\gamma}[1-a(1-q)x_1^{\delta}]^{\frac{\eta}{1-q}} \tag{2.32}$$

for $q<1, \eta>0, a>0, \delta>0$ where

$$c_{11} = \frac{\delta[a(1-q)]^{\frac{\gamma+1}{\delta}}\Gamma(\frac{\gamma+1}{\delta}+\frac{\eta}{1-q}+1)}{\Gamma(\frac{\gamma+1}{\delta})\Gamma(\frac{\eta}{1-q}+1)}.$$

Then the density of $u_2 = x_1x_2$, when x_1 has the above density $f_{11}(x_1)$ and x_2 has arbitrary density $f(x_2)$, is denoted by $g_{21}(u_2)$, and it is given by

$$\begin{aligned} g_{21}(u_2) &= c_{11}\int_v \frac{1}{v} f_1(\frac{u_2}{v}) f_2(v)\mathrm{d}v \\ &= c_{11}\int_v \frac{1}{v}(\frac{u_2}{v})^{\gamma}[1-a(1-q)(\frac{u_2}{v})^{\delta}]^{\frac{\eta}{1-q}} f(v)\mathrm{d}v \\ &= c_{11}u_2^{\gamma}\int_{v=u_2[a(1-q)]^{\frac{1}{\delta}}}^{\infty} \{[v^{\delta}-a(1-q)u_2^{\delta}]^{\frac{\eta}{1-q}}\, v^{-\gamma-(\frac{\eta\delta}{1-q}+1)}\} f(v)\mathrm{d}v \end{aligned} \tag{2.33}$$

Some particular cases here will be interesting.

2.11 Special Cases

Case (1): When $\delta = m, q = 0, \frac{\eta}{1-q} = \alpha - 1$ then (2.33) is the result (2.6.9) of Mathai and Haubold [18].

Case (2): When $\delta = 1, a = 1, q = 0, \eta = \alpha - 1$ then

$$\frac{\Gamma(\gamma+1)}{\Gamma(\gamma+1+\alpha)}g_{21}(u_2)=\frac{1}{\Gamma(\alpha)}u^{\gamma}\int_{v=u_2}^{\infty}(v-u_2)^{\alpha-1}v^{-\gamma-\alpha}f(v)\mathrm{d}v == K_{2,u_2,\gamma}^{-\alpha}f \tag{2.34}$$

is Erdélyi-Kober fractional integral of the second kind.

Case (3): When $\delta = 1, a = 1, q = 0, \eta = \alpha - 1, \gamma = 0$ then

$$\begin{aligned}\frac{1}{\Gamma(\alpha+1)}g_{21}(u_2) &= \frac{1}{\Gamma(\alpha)}\int_{t=u_2}^{\infty}(t-u_2)^{\alpha-1}t^{-\alpha}f(t)\mathrm{d}t \\ &= K_{2,u_2,0}^{-\alpha}f = {}_xW_{\infty}^{-\alpha}t^{-\alpha}f(t) = W_{2,x}^{-\alpha}t^{-\alpha}f(t)\end{aligned} \tag{2.35}$$

where ${}_xW_{\infty}^{-\alpha}f = W_{2,x}^{-\alpha}f$ is the Weyl right-sided fractional integral of order α, which can also be called the right-sided Riemann-Liouville fractional integral of order α when the right limit is at ∞, for the function $t^{-\alpha}f(t)$.

Let us look at (2.33) when q varies from $-\infty$ to 1. Here (2.33) is a collection of generalized Erdélyi-Kober operators of the second kind operating on f. It can also be considered as a Mellin convolution of a product where one function $f(x_2)$ is arbitrary and the other function $f_{11}(x_1)$ is of the form in (2.32). Here q describes a path of movement of the Erdélyi-Kober fractional integral of the second kind. In the limit when $q \to 1_-$ then (2.33) will be

$$\lim_{q\to 1_-} g_{21}(u_2)=c_{13}\int_{v=0}^{\infty}\frac{1}{v}(\frac{u}{v})^{\gamma}\mathrm{e}^{-a\eta(\frac{u_2}{v})^{\delta}}f(v)\mathrm{d}v=c_{13}u_2^{\gamma}\int_{v=0}^{\infty}v^{-\gamma-1}\mathrm{e}^{-a\eta(\frac{u_2}{v})^{\delta}}f(v)\mathrm{d}v \tag{2.36}$$

where

$$c_{13} = \delta\frac{(a\eta)^{\frac{\gamma+1}{\delta}}}{\Gamma(\frac{\gamma+1}{\delta})}.$$

This (2.36) is also connected to Krätzel transform if $f(v)$ can be written as $\mathrm{e}^{-bv}\psi(v)$. Then (2.36) will correspond to generalized Krätzel transform of $\psi(v)$. Krätzel transform is widely applied in various disciplines. This Krätzel transform is also connected to inverse Gaussian density in stochastic processes, to Bayesian analysis, to reaction rate probability integral in reaction rate theory, to P-transforms, to fractional integral transforms and many other topics, the details may be seen from Kumar, Kumar and Haubold, Kumar and Kilbas, Mathai, and Mathai, Provost and Hayakawa [8–10, 15, 20].

We can also consider the case when $q > 1$. Writing $1 - q = -(q - 1)$ with $q > 1$, $f_{11}(x_1)$ of (2.32) changes to the following form.

$$f_{12}(x_1) = c_{12} x_1^{\gamma} [1 + a(q-1)x_1^{\delta}]^{-\frac{\eta}{q-1}}, a > 0, \delta > 0, \eta > 0, q > 1 \tag{2.37}$$

where

$$c_{12} = \delta \frac{[a(q-1)]^{\frac{\gamma+1}{\delta}} \Gamma(\frac{\eta}{q-1})}{\Gamma(\frac{\gamma+1}{\delta})\Gamma(\frac{\eta}{q-1} - \frac{\gamma+1}{\delta})}, \quad \frac{\eta}{q-1} - \frac{\gamma+1}{\delta} > 0$$

and

$$c_{12} \to \delta \frac{(a\eta)^{\frac{\gamma+1}{\delta}}}{\Gamma(\frac{\gamma+1}{\delta})} = c_{13} \text{ when } q \to 1_+.$$

In this case $0 < x_1 < \infty$. Then the density of the product $u_2 = x_1 x_2$, denoted as $g_{22}(u_2)$, is the following:

$$\begin{aligned} g_{22}(u) &= c_{12} \int_{t=0}^{\infty} \frac{1}{t} (\frac{u_2}{t})^{\gamma} [1 + a(q-1)(\frac{u_2}{t})^{\delta}]^{-\frac{\eta}{q-1}} f(t) \mathrm{d}t \\ &= c_{12} u_2^{\gamma} \int_0^{\infty} t^{-\gamma + (\frac{\delta\eta}{q-1} - 1)} [t^{\delta} + a(q-1)u_2^{\delta}]^{-\frac{\eta}{q-1}} f(t) \mathrm{d}t. \end{aligned} \tag{2.38}$$

This can also be taken as a generalization of the Erdélyi-Kober fractional integral of the second kind. When $q \to 1_+$, (2.38) goes into the following form.

$$\lim_{q \to 1_+} g_{23}(u_2) = c_{13} \int_0^{\infty} t^{-\gamma-1} \mathrm{e}^{-a\eta(\frac{u}{t})^{\delta}} f(t) \mathrm{d}t \tag{2.39}$$

It is easy to note that

$$\lim_{q \to 1_-} c_{11} = \lim_{q \to 1_+} c_{12} = c_{13}.$$

2.12 Another Form of Generalization of Erdélyi-Kober Operators of the Second Kind

One can take any specific density for $f_5(x_1)$ and an arbitrary density for $f_6(x_2)$. Then take the Mellin convolution of a product. Then this will give a class of generalized Erdélyi-Kober operators from a statistical point of view. For a fractional integral, which is found in the literature, one needs a type-1 beta form for $f_5(x_1)$. We can also relocate x_1 at $x_1 = b$ so that $x_1 \geq b$ for some b. Then also we can obtain a fractional integral. From a mathematical point of view such a generalization may not have much of a significance.

Another generalization is available in terms of hypergeometric functions. We can append a convergent series to any given density, for example, append a hypergeometric series to $f_5(x_1)$, to get a general form. In model building situations such appended forms may produce thicker or thinner tails and hence useful in model building. Let us consider appending a hypergeometric series to the basic density $f_5(x_1)$ of x_1. Let the appended density be denoted by f_7.

$$f_7(x_1) = \frac{1}{c_7}\ {}_pF_q(a_1, \ldots, a_p; b_1, \ldots, b_q; ax_1)x_1^{\beta}(1 - x_1)^{\alpha-1},\ 0 < x_1 < 1 \tag{2.40}$$

and $f_7(x_1) = 0$ elsewhere, where c_7^{-1} is the normalizing constant. We can create a statistical density out of this form as follows: In order to assure nonnegativity of the function let us assume that the parameters $a_1, \ldots, a_p, b_1, \ldots, b_q, a$ are all positive. Then, for $q \geq p$ or $p = q + 1$ and $|ax| < 1$,

$$\begin{aligned}&{}_pF_q(a_1, \ldots, a_p; b_1, \ldots, b_q : ax_1)x_1^{\beta}(1 - x_1)^{\alpha-1}\\ &= \sum_{k=0}^{\infty} \frac{(a_1)_k \ldots (a_p)_k}{(b_1)_k \ldots (b_q)_k} \frac{a^k x_1^k}{k!} x_1^{\beta}(1 - x_1)^{\alpha-1},\end{aligned}$$

where, for example, $(a)_k = a(a + 1) \ldots (a + k - 1), a \neq 0, (a)_0 = 1$ is the Pochhammer symbol . Total integral is available from the basic type-1 beta integral

$$\int_0^1 x_1^{\beta+k}(1 - x_1)^{\alpha-1}\mathrm{d}x_1 = \frac{\Gamma(\beta + 1 + k)\Gamma(\alpha)}{\Gamma(\alpha + \beta + 1 + k)} = \frac{\Gamma(\alpha)\Gamma(\beta + 1)}{\Gamma(\alpha + \beta + 1)} \frac{(\beta + 1)_k}{(\alpha + \beta + 1)_k}.$$

The normalizing constant is $\frac{1}{c_7}$ where

$$c_7 = \frac{\Gamma(\alpha)\Gamma(\beta + 1)}{\Gamma(\alpha + \beta + 1)}\ {}_{p+1}F_{q+1}(a_1, \ldots, a_p, \beta + 1; b_1, \ldots, b_q, \alpha + \beta + 1; a),$$

for $\Re(\alpha) > 0, \Re(\beta) > 0$. Then

$$f_7(x_1) = \frac{1}{c_7}\ {}_pF_q(a_1, \ldots, a_p; b_1, \ldots, b_q; ax_1)x_1^{\beta}(1 - x_1)^{\alpha-1}, 0 < x_1 < 1$$

and zero elsewhere is a density. Take this form of $f_7(x_1)$ and $f_8(x_2) = f(x_2)$ an arbitrary density, and proceed to find the density of $u_2 = x_1x_2$ as before. Denoting the density by $g_{24}(u_2)$, we have

$$g_{24}(u_2) = \frac{1}{c_7}\sum_{k=0}^{\infty}\frac{(a_1)_k\ldots(a_p)_k}{(b_1)_k\ldots(b_q)_k}\frac{a^k}{k!}\int_v u_2^{\beta}(v-u_2)^{\alpha-1}v^{-\beta-\alpha}(\frac{u_2}{v})^k f(v)\mathrm{d}v$$
$$= \frac{1}{c_7}u_2^{\beta}\int_{v>u_2>0}(v-u_2)^{\alpha-1}v^{-\beta-\alpha}\,{}_pF_q(a_1,\ldots,a_p;b_1,\ldots,b_q;\frac{au_2}{v})f(v)\mathrm{d}v. \tag{2.41}$$

A particular case of this for a ${}_2F_1$ is equation (2.7.2) of Mathai and Haubold [18]. This particular case was given by others earlier. There is a serious drawback in taking a ${}_2F_1$ because there may be problems in taking Laplace, Mellin and other transforms for the convergence of the series forms. It is safer to take $q \geq p$ in the case of appending a hypergeometric series to the type-1 beta form for $f_7(x_1)$. Note that (2.41) is a generalization of Erdélyi-Kober operator of the second kind operating on f as well as one has an interpretation in terms of a statistical density.

Another form of appending a hypergeometric series is to consider a hypergeometric series with argument $a(1-x_1)$ instead ax_1. Going through the same process as before, one can create a statistical density of the form, denoted by $f_9(x_1)$,

$$f_9(x_1) = \frac{1}{c_9}\sum_{k=0}^{\infty}\frac{(a_1)_k\ldots(a_p)_k}{(b_1)_k\ldots(b_q)_k}\frac{a^k}{k!}x_1^{\beta}(1-x_1)^{\alpha-1+k} \tag{2.42}$$

for $0 < x_1 < 1$ and zero elsewhere, where

$$c_9 = \frac{\Gamma(\beta+1)\Gamma(\alpha)}{\Gamma(\alpha+\beta+1)}\,{}_{p+1}F_{q+1}(a_1,\ldots,a_p,\alpha;b_1,\ldots,b_q,\beta+1+\alpha;a),$$

fpr $\Re(\alpha) > 0, \Re(\beta) > 0$. In order to guarantee nonnegativity we may assume all parameters a_j's, b_j's, be positive, $a > 0$, $\alpha > 0, q \geq p$. If $p = q+1$ then take $|a(1-x_1)| < 1$. Proceeding exactly as before, taking x_1 having this appended density $f_9(x_1)$ and x_2 having an arbitrary density $f_{10}(x_2) = f(x_2)$, then the density of $u_2 = x_1x_2$, denoted by $g_{25}(u_2)$, is available as

$$g_{25}(u_2) = \frac{1}{c_9}\sum_{k=0}^{\infty}\frac{(a_1)_k\ldots(a_p)_k}{(b_1)_k\ldots(b_q)_k}\frac{a^k}{k!}\int_v\frac{1}{v}(\frac{u_2}{v})^{\beta}(1-\frac{u_2}{v})^{\alpha+k-1}f(v)\mathrm{d}v.$$

This can be simplified to the form

$$g_{25}(u_2) = \frac{u_2^{\beta}}{c_9}\int_{v>u_2>0}(v-u_2)^{\alpha-1}v^{-\beta-\alpha}\,{}_pF_q(a_1,\ldots,a_p;b_1,\ldots,b_q;a(1-\frac{u_2}{v}))f(v)\mathrm{d}v, \tag{2.43}$$

for $|a| < 1, v > u_2 > 0$. This is a generalization of Erdélyi-Kober fractional integral of the second kind. Here also, one could have taken the argument of ${}_pF_q$ as $a^{\delta_1}(1-x_1)^{\delta_2}$. These will provide more generalized forms. A particular case of (2.43) is (2.7.4) of Mathai and Haubold [18]. This particular case, in terms of a ${}_2F_1$, was given by others earlier. As remarked above, there is a disadvantage in taking a ${}_2F_1$. This special case in terms of a ${}_2F_1$ is Saigo operator, see (2.7.8) of Mathai and Haubold [18].

Remark 2.1 It may be noted from (2.41) and (2.43) that one can consider $f_9(x_1)$ in terms of a hypergeometric function with argument ax_1 or $a^{\delta_1}x_1^{\delta_2}$ or $a(1-x_1)$ or $a^{\delta_1}(1-x_1)^{\delta_2}$ or $a^{\delta_1}(1-x_1)^{\delta_2}x_1^{\delta_3}$ with $\delta_j > 0, j = 1, 2, 3$. The procedure will be the same. If statistical densities are not needed then one can take multiplicative factors for $f_9(x_1)$ as well as for $f_{10}(x_2)$. Instead of a hypergeometric series, one can consider $f_9(x_1)$ in terms of a Meijer's G-function or H-function with arguments any one of the forms mentioned above. If $g_2(u_2)$ to remain as a statistical density then, apart from convergence of the series and integrals, the parameters are to be restricted so that the functions remain positive in the range $0 < x_1 < 1$ and zero outside this range. Since these generalizations are routine mathematical exercises, we will not give the explicit expressions for each generalization of Erdélyi-Kober fractional integral of the second kind here.

2.13 Mellin Transform of the Generalized Erdélyi-Kober Operator of the Second Kind

For the generalized form in (2.41) the Mellin transform is available by evaluating the integral, through interchange of integrals,

$$\begin{aligned}
&\int_0^\infty u_2^{s-1}u_2^{\beta+k}[\int_{v>u_2>0}(v-u_2)^{\alpha-1}v^{-\beta-\alpha-k}f(v)\mathrm{d}v]\\
&=\int_{v=0}^\infty v^{-\beta-\alpha-k}f(v)[\int_{u_2=0}^v u_2^{s-1+\beta+k}(v-u_2)^{\alpha-1}\mathrm{d}u_2]\mathrm{d}v\\
&=\int_{v=0}^\infty v^{s-1}f(v)\mathrm{d}v\int_0^1 y^{s+\beta+k-1}(1-y)^{\alpha-1}\mathrm{d}y\\
&=\frac{\Gamma(\alpha)\Gamma(\beta+s)}{\Gamma(\alpha+\beta+s)}\frac{(\beta+s)_k}{(\alpha+\beta+s)_k}f^*(s),
\end{aligned}$$

for $\Re(\alpha) > 0, \Re(\beta+s) > 0$. Therefore the Mellin transform of (2.41) is the following:

$$\begin{aligned}
&M\{g_{24}(u_2)\text{ of (2.41)}; s\}\\
&=\frac{\Gamma(\alpha)}{c_9}\frac{\Gamma(\beta+s)}{\Gamma(\alpha+\beta+s)}{}_{p+1}F_{q+1}(a_1,\ldots,a_p,\beta+s;b_1,\ldots,b_q,\alpha+\beta+s;a).
\end{aligned} \tag{2.44}$$

In a similar manner one can compute the Mellin transform of $g_{25}(u)$ of (2.43). The basic integral to be evaluated is the following, through interchange of integrals:

$$\begin{aligned}
&\int_{u_2=0}^{\infty} u_2^{s-1+\beta}[\int_{v=u_2>0}^{\infty}(v-u_2)^{\alpha-1}v^{-\beta-\alpha}(1-\frac{u_2}{v})^k \mathrm{d}u_2]f(v)\mathrm{d}v\\
&=\int_{v=0}^{\infty} v^{-\beta-\alpha+\alpha-1}f(v)[\int_{u_2=0}^{v}(1-\frac{u_2}{v})^{\alpha+k-1}u_2^{\beta+s-1}\mathrm{d}u_2]\mathrm{d}v\\
&=\int_{v=0}^{\infty} v^{s-1}f(v)\mathrm{d}v[\int_0^1 y^{\beta+s-1}(1-y)^{\alpha+k-1}\mathrm{d}y]\\
&=f^*(s)\frac{\Gamma(\beta+s)\Gamma(\alpha)}{\Gamma(\alpha+\beta+s)}\frac{(\alpha)_k}{(\alpha+\beta+s)_k}, \Re(\alpha)>0, \Re(\beta+s)>0.
\end{aligned}$$

Hence

$$M\{g_{25}(u_2) \text{ of } (2.43); s\} = \frac{\Gamma(\alpha)}{c_9}\frac{\Gamma(\beta+s)}{\Gamma(\alpha+\beta+s)}{}_{p+1}F_{q+1}(a_1,\ldots,a_p,\alpha;b_1,\ldots,b_q,\alpha+\beta+s;a). \tag{2.45}$$

The right-sided Weyl fractional integral of order α is given by

$$ {}_xW_{\infty}^{-\alpha}f = W_{2,x}^{-\alpha}f = \frac{1}{\Gamma(\alpha)}\int_x^{\infty}(t-x)^{\alpha-1}f(t)\mathrm{d}t, \Re(\alpha)>0. \tag{2.46}$$

The Mellin transform is $\frac{\Gamma(s)}{\Gamma(\alpha+s)}f^*(\alpha+s)$. Such a form can be generated from the Mellin convolution of a product as well as from a statistical density. Consider

$$f_{15}(x_1) = \frac{1}{\Gamma(\alpha)}(1-x_1)^{\alpha-1}, \Re(\alpha)>0.$$

Note that $f_{15}(x_1)$ here is a constant multiple of a statistical density. In fact, $\Gamma(\alpha+1)f_{15}(x_1)$ is a type-1 beta density. Let $f_{16}(x_2)=f(x_2)$ be an arbitrary function. Let $u_2=x_1x_2, v=x_1, x_2=\frac{v}{u_2}$. Then the Mellin convolution of a product for $f(x_2)$ and $x_1^{-\alpha}f_{15}(x_1)$ is given by the following:

$$\begin{aligned}
&\int_v \frac{v^{-\alpha}}{\Gamma(\alpha)}(1-v)^{\alpha-1}f(\frac{u_2}{v})\frac{1}{v}\mathrm{d}v, (t=\frac{u_2}{v}, \mathrm{d}v=-\frac{u_2}{t^2}\mathrm{d}t)\\
&=\frac{1}{\Gamma(\alpha)}\int_t \frac{1}{t}(\frac{u_2}{t})^{-\alpha}(1-\frac{u_2}{t})^{\alpha-1}f(t)\mathrm{d}t\\
&=\frac{u_2^{-\alpha}}{\Gamma(\alpha)}\int_{t=u_2}^{\infty}(t-u_2)^{\alpha-1}f(t)\mathrm{d}t = u_2^{-\alpha}[{}_xW_{\infty}^{-\alpha}f] = u_2^{-\alpha}W_{2,x}^{-\alpha}f.
\end{aligned} \tag{2.47}$$

Theorem 2.4 *The right-sided Weyl fractional integral operator of order* α, ${}_xW_{\infty}^{-\alpha} = W_{2,x}^{-\alpha}$, *operating on* f, *can be taken as a Mellin convolution of a product of* $f_{15}(x_1)$ *and* $f_{16}(x_2) = f(x_2)$ *where* $f_{15}(x_1)$ *is a constant multiple of a type-1 beta density.*

2.14 A Geometrical and Some Physical Interpretations of Fractional Integrals

The name "fractional integral" suggests any type of incomplete integral. In this respect an incomplete gamma function or incomplete beta function or any integral of the form $\int_{-\infty}^{a} f(x)\mathrm{d}x$ or of the form $\int_{a}^{\infty} f(x)\mathrm{d}x$ or $\int_{0}^{a} f(x)\mathrm{d}x$ or $\int_{a}^{b} f(x)\mathrm{d}x$ could be taken as fractions of total integrals. But in the literature of fractional integrals, any such fraction of the total integral is not taken as "fractional integrals". It may be noted that a certain fraction of the total integral is taken but the structure is that it is a fraction of the total integral in a product of two functions where one is a type-1 beta type. Consider the product of two real-valued scalar functions of the real scalar variables x_1 and x_2, in the form $f_1(x_1)f_2(x_2)$. If one function is a power function of the form $f_1(x_1) = x_1^{\alpha-1}$ and if we take the Laplace convolution for a sum, denoted by $D_{f_2}^{-\alpha}$, then

$$D_{f_2}^{-\alpha} = \int_{t=a}^{x} (x-t)^{\alpha-1} f_2(t)\mathrm{d}t \tag{2.48}$$

for some a, including $a = -\infty$. If both x_1 and x_2 are restricted to be positive variables then $a = 0$. In the Laplace convolution of a sum we are making the transformation $x = x_1 + x_2, x_2 = t$. This integral in (2.48), divided by a constant $\Gamma(\alpha), \Re(\alpha) > 0$, is the left-sided or first kind Riemann-Liouville fractional integral in the literature. Hence one interpretation is that Riemann-Liouville left-sided fractional integral is the Laplace convolution for a sum where one function is a power function.

2.14.1 An Interpretation in Terms of Densities of Sum and Difference

Let us consider two real scalar positive random variables $x_1 > 0, x_2 > 0$. Let the densities be $f_1(x_1)$ and $f_2(x_2)$ respectively. Let the variables be independently distributed or enjoy the product probability property (ppp). Then the joint density is $f_1(x_1)f_2(x_2)$. Let $z_1 = x_1 + x_2$ and $z_2 = x_2 - x_1$, $x_2 = t$, with $z_2 \geq 0$. Then the density of z_1, denoted by $h_1(z_1)$, is the following, when $f_1(x_1) = c_1 x_1^{\alpha-1}$,

$0 \le x_1 \le 1$, $f_1(x_1) = 0$ elsewhere, a power function density, where c_1 is the normalizing constant:

$$h_1(z_1) = c_1 \int_0^{z_1} (z_1 - t)^{\alpha-1} f_2(t)\mathrm{d}t. \tag{2.49}$$

A constant multiple in (2.49) or if c_1 is replaced by $\frac{1}{\Gamma(\alpha)}$, $\Re(\alpha) > 0$ then (2.49) is Riemann-Liouville first kind or left-sided fractional integral of order α. Hence this fractional integral can be interpreted as a constant multiple of a density of a sum. Now, let us look into the density of the difference z_2 when $f_1(x_1)$ is again a power function density as above. Then the density of z_2, denoted by $h_2(z_2)$, is given by the following:

$$h_2(z_2) = c_1 \int_{z_2}^{\infty} (t - z_2)^{\alpha-1} f_2(t)\mathrm{d}t. \tag{2.50}$$

If c_1 is replaced by $\frac{1}{\Gamma(\alpha)}$, $\Re(\alpha) > 0$ then we obtain Riemann-Liouville right-sided or second kind fractional integral of order α. Therefore, a constant multiple of (2.50) is Riemann-Liouville right-sided fractional integral of order α.

2.14.2 *Fractional Integrals as Fractions of Total Probabilities*

Consider the total integral coming from a type-1 beta density with parameters (β, α), that is, observing that the total integral in a statistical density is 1,

$$\begin{aligned} 1 &= \frac{\Gamma(\alpha+\beta)}{\Gamma(\alpha)\Gamma(\beta)} \int_0^1 y^{\beta-1}(1-y)^{\alpha-1}\mathrm{d}y \\ &= \frac{\Gamma(\alpha+\beta)}{\Gamma(\alpha)\Gamma(\beta)} \int_0^x (\frac{t}{x})^{\beta-1}(1-\frac{t}{x})^{\alpha-1}\mathrm{d}t \\ x^{\alpha+\beta-1}(1) &= \frac{\Gamma(\alpha+\beta)}{\Gamma(\alpha)\Gamma(\beta)} \int_0^x (x-t)^{\alpha-1} t^{\beta-1}\mathrm{d}t. \end{aligned} \tag{2.51}$$

The left-side in the last line of (2.51) is a fraction of the total probability 1, namely $x^{\alpha+\beta-1}$ times (1) observing that $0 < x < 1$. Hence (2.51), which is a Riemann-Liouville left-sided fractional integral of order α where the arbitrary function is $t^{\beta-1}$, is a fraction of the total probability in a statistical density.

Consider the total probability coming from a gamma density with parameter α. Then we have

$$1 = \frac{1}{\Gamma(\alpha)} \int_0^{\infty} y^{\alpha-1}\mathrm{e}^{-y}\mathrm{d}y, \Re(\alpha) > 0.$$

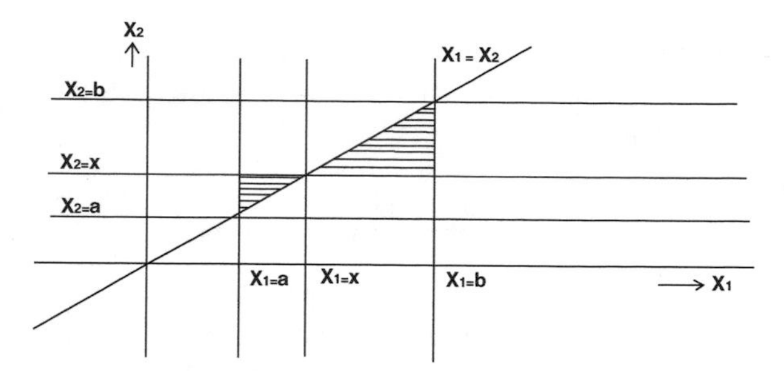

Fig. 2.1 Integration over simplices

Consider a fraction of this total probability, namely $e^{-x}(1)$. Then

$$
\begin{aligned}
e^{-x}(1) &= \frac{e^{-x}}{\Gamma(\alpha)}\int_0^{\infty} y^{\alpha-1}e^{-y}dy \\
&= \frac{1}{\Gamma(\alpha)}\int_0^{\infty} y^{\alpha-1}e^{-(x+y)}dy, \ (t = x + y) \\
&= \frac{1}{\Gamma(\alpha)}\int_x^{\infty} (t-x)^{\alpha-1}e^{-t}dt. \qquad (2.52)
\end{aligned}
$$

The right side in the last line of (2.52) is right-sided Riemann-Liouville fractional integral of order α where the arbitrary function is e^{-t} and the left side is a fraction of a total probability.

2.14.3 A Geometrical Interpretation

Here we consider a geometrical interpretation of the left-sided as well as the right-sided Riemann-Liouville fractional integral of order α. The following geometrical interpretation is given by Mathai [17]. Consider the following square of length $b-a$, the line $x_1 = x_2$ and the left-sided and right-sided triangles as shown in Fig. 2.1.

Consider the area of the triangle on the left of the line $x_1 = x_2$ in a plane or 2-space. We will use physicist's notation of writing the differential elements soon after the integral sign, for convenience. The area

$$\int_{x_1=a}^{x} dx_1 \int_{x_2=x_1}^{x} dx_2 = \int_{x_1=a}^{x} (x-x_1)dx_1.$$

But for the corresponding simplex in 3-space the volume is given by the following:

$$\int_{x_1=a}^{x} dx_1 \int_{x_2=x_1}^{x} dx_2 \int_{x_3=x_2}^{x} dx_3 = \int_{x_1=a}^{x} x_1 \int_{x_2=x_1}^{x} dx_2(x-x_2) = \int_{x_1=a}^{x} \frac{(x-x_1)^2}{2!}dx_1.$$

Then the volume of the left-sided simplex in the n-space is $= \int_{x_1=a}^{x} \frac{(x-x_1)^{n-1}}{(n-1)!}\mathrm{d}x_1$.. Note that if we had integrated out a function of x_1 alone, say $f(x_1)$ the result would have been

$$\int_{x_1=a}^{x} \frac{(x-x_1)^{n-1}}{(n-1)!} f(x_1)\mathrm{d}x_1 = \int_{t=a}^{x} \frac{(x-t)^{n-1}}{\Gamma(n)} f(t)\mathrm{d}t.$$

We may denote this left-sided integral as ${}_aD_x^{-n} f(t) = D_{1,(a,x)}^{-n} f$ the integral over the left simplex in n-space. If n is replaced by a general α then we have the left-sided or first kind Riemann-Liouville fractional integral of order α, denoted by

$${}_aD_x^{-\alpha} f(t) = D_{1,(a,x)}^{-\alpha} f = \frac{1}{\Gamma(\alpha)} \int_{t=a}^{x} (x-t)^{\alpha-1} f(t)\mathrm{d}t, \Re(\alpha) > 0,$$

and hence the corresponding fractional derivative of order α will be denoted by ${}_aD_x^{\alpha} f(t) = D_{1,(a,x)}^{\alpha} f$. Here $-\alpha$ indicates integral and $+\alpha$ indicates derivative.

Now, consider the right-sided triangle in Fig. 2.1.Its area is given by

$$\int_{x_1=x}^{b} \mathrm{d}x_1 \int_{x_2=x}^{x_1} \mathrm{d}x_2 = \int_{x_1=x}^{b} (x_1 - x)\mathrm{d}x_1.$$

The volume of a corresponding simplex in 3-space is

$$\int_{x_1=x}^{b} \mathrm{d}x_1 \int_{x_2=x}^{x_1} \mathrm{d}x_2 \int_{x_3=x}^{x_2} \mathrm{d}x_3 = \int_{x_1=x}^{b} \frac{(x_1-x)^2}{2!}\mathrm{d}x_1.$$

The volume of the right-sided simplex in n-space is then

$$\int_{x_1=x}^{b} \frac{(x_1-x)^{n-1}}{(n-1)!}\mathrm{d}x_1 = \int_{x_1=x}^{b} \frac{(x_1-x)^{n-1}}{\Gamma(n)}\mathrm{d}x_1.$$

Now, if an arbitrary function of x_1 alone is integrated over this simplex then the volume over the simplex, which is a fraction of the volume over the cube, is given by

$${}_xD_b^{-n} f(t) = D_{2,(x,b)}^{-n} f = \int_{x_1=x}^{b} \frac{(x_1-x)^{n-1}}{\Gamma(n)} f(x_1)\mathrm{d}x_1 = \int_{x}^{b} \frac{(t-x)^{n-1}}{\Gamma(n)} f(t)\mathrm{d}t.$$

If n is replaced by an arbitrary α then the right-sided integral of order α, denoted by ${}_xD_b^{-\alpha} f(t) = D_{2,(x,b)}^{-\alpha} f$ is given by

$${}_xD_b^{-\alpha} f(t) = D_{2,(x,b)}^{-\alpha} f = \frac{1}{\Gamma(\alpha)} \int_{t=x}^{b} (t-x)^{\alpha-1} f(t)\mathrm{d}t, \Re(\alpha) > 0.$$

This is the Riemann-Liouville right-sided or second kind fractional integral of order α. The corresponding derivative of order α will be denoted by ${}_xD_b^{\alpha} = D_{2,(x,b)}^{\alpha} f$, taking derivative as anti-integral.

2.15 A General Definition of Fractional Integrals

From all the discussions and results so far, we have seen the following: Fractional integrals considered in the literature are not arbitrary type of incomplete integrals. They are incomplete integrals coming from a product structure where one function has a part which is of the form of a type-1 beta function and the other function has a part which is an arbitrary function. This is the common feature in all the fractional integrals in the literature, originally coming from a consideration of the geometry given in Fig. 2.1. Hence one can have a general definition based on Mellin convolutions of products and ratios or densities of products and ratios. Let our original $f_1(x_1)$ and $f_2(x_2)$ be of the following forms:

$$f_1(x_1) = \frac{1}{\Gamma(\alpha)}\phi_1(x_1)(1-x_1)^{\alpha-1} \text{ and } f_2(x_2) = \phi_2(x_2)f(x_2) \tag{2.53}$$

where $\Re(\alpha) > 0$, ϕ_1 and ϕ_2 are specified functions, f is an arbitrary function. From the structure in (2.53), where f_1 has one part a specified function and the other part a type-1 beta function, f_2 has one part a specified function and the other part an arbitrary function. By specifying ϕ_1 and ϕ_2 and then taking Mellin convolutions of ratios and products we can derive all the left-sided and right-sided fractional integrals in the literature.

2.15.1 Mellin Convolution of Product and Second Kind Integrals

Let us consider the Mellin convolution of a product, denoted by $g_2(u_2)$, where $u_2 = x_1x_2$, $v = x_2$ or $x_1 = \frac{u_2}{v}$, $x_2 = v$, then

$$g_2(u_2) = \frac{1}{\Gamma(\alpha)}\int_v \frac{1}{v}\phi_1(\frac{u_2}{v})(1-\frac{u_2}{v})^{\alpha-1}\phi_2(v)f(v)\mathrm{d}v, \Re(\alpha) > 0. \tag{2.54}$$

In (2.54) let $\phi_1 = 1$ and $\phi_2(v) = v^{\alpha}$. Then (2.54) becomes

$$g_2(u_2) = \frac{1}{\Gamma(\alpha)}\int_{v=u_2}^{b} (v-u_2)^{\alpha-1}f(v)\mathrm{d}v, \Re(\alpha) > 0. \tag{2.55}$$

This is Riemann-Liouville second kind or right-sided fractional integral of order α, namely $D_{2,(u_2,b)}^{-\alpha}f$ if there is an upper bound for v. If the upper bound is $+\infty$ then (2.55) is second kind or right-sided Weyl fractional integral of order α.

Let $\phi_1(x_1) = x_1^{\gamma}, \phi_2 = 1$ in (2.54) and let the upper bound for v be $+\infty$. Then (2.54) becomes the following, again denoted by $g_2(u_2)$:

$$g_2(u_2) = \frac{u_2^{\gamma}}{\Gamma(\alpha)} \int_{v=u_2}^{\infty} v^{-\gamma-\alpha}(v-u_2)^{\alpha-1} f(v)\mathrm{d}v, \Re(\alpha) > 0. \tag{2.56}$$

This (2.56) is nothing but Erdélyi-Kober fractional integral of the second kind of order α and parameter γ, namely, $K_{2,u_2,\gamma}^{-\alpha}f$.

2.15.2 Mellin Convolution of a Ratio and First Kind Fractional Integrals

Let $u_1 = \frac{x_2}{x_1}, x_2 = v$ or $x_1 = \frac{v}{u_1}$ and the Jacobian is $-\frac{v}{u_1^2}$. Again, let $f_1(x_1)$ and $f_2(x_2)$ be as in (2.53). Then the Mellin convolution of a ratio or the density of a ratio when f_1 and f_2 are statistical densities, denoted by $g_1(u_1)$, is the following:

$$g_1(u_1) = \frac{1}{\Gamma(\alpha)} \int_v \frac{v}{u_1^2}\phi_1(\frac{v}{u_1})(1-\frac{v}{u_1})^{\alpha-1}\phi_2(v)f(v)\mathrm{d}v. \tag{2.57}$$

Now, by specializing ϕ_1 and ϕ_2 one should get all the left-sided or first kind fractional integrals in the literature from (2.57).

Let $\phi_1(x_1) = x_1^{\gamma-1}, \phi_2 = 1$ in (2.57). Then (2.57) becomes the following, again denoted by $g_1(u_1)$:

$$g_1(u_1) = \frac{u_1^{-\gamma-\alpha}}{\Gamma(\alpha)} \int_{v=0}^{u_1} v^{\gamma}(u_1-v)^{\alpha-1} f(v)\mathrm{d}v. \tag{2.58}$$

This is Erdélyi-Kober fractional integral of the first kind of order α and parameter γ, namely $K_{1,u_1,\gamma}^{-\alpha}f$. Let $\phi_1(x_1) = x_1^{-\alpha-1}, \phi_2(v) = v^{\alpha}$. Then (2.57) becomes the following, again denoted by $g_1(u_1)$:

$$g_1(u_1) = \frac{1}{\Gamma(\alpha)} \int_{v=0}^{u_1} (u_1-v)^{\alpha-1} f(v)\mathrm{d}v, \Re(\alpha) > 0. \tag{2.59}$$

This (2.59) is Riemann-Liouville left-sided or first kind fractional integral of order α where the left limit is zero, namely, $D_{1,u_1}^{-\alpha}f$ or ${}_0D_{u_1}{}^{-\alpha}f$.

References

1. G. Calcagni, Geometry of fractional spaces. Adv. Theor. Math. Phys. **16**, 549–644 (2012)
2. R. Gorenflo, F. Mainardi, Fractional calculus integral and differential equations of fractional order, in *Fractal and Fractional Calculus in Continuum Mechanics*, ed. by A. Carpinteri, F. Mainardi (Springer, Wien and New York, 1997), pp. 223–276
3. R. Herrmann, Towards a geometric interpretation of generalized fractional integrals – Erdélyi-Kober type integrals on R(N) as an example. Fract. Calc. Appl. Anal. **17(2)**, 361–370 (2014)
4. R.Hilfer (ed.), *Applications of Fractional Calculus in Physics* (World Scientific, Singapore, 2000)
5. A.A. Kilbas, J.J. Trujillo, Computation of fractional integral via functions of hypergeometric and Bessel type. J. Comput. Appl. Math. **118**(1–2), 223–239 (2000)
6. V. Kiryakova, *Generalized Fractional Calculus and Applications*, Pitman Res Notes Math 301, Longman Scientific & Technical: Harlow, Co-published with (Wiley, New York, 1994)
7. V. Kiryakova, Y. Luchko, Riemann-Liouville and Caputo type multiple Erdélyi-Kober operators. Centr. Eur. J. Phys. **11(10)**, 1314–1336 (2013)
8. D. Kumar, P-transform. Integral Transforms Spec. Funct. **22(8)**, 603–616 (2011)
9. D. Kumar, H.J. Haubold, On extended thermonuclear functions through pathway model. Adv. Space Res. **45**, 698–708 (2010)
10. D. Kumar, A.A. Kilbas, Fractional calculus of P-transform. Fract. Calc. Appl. Anal. **13(3)**, 309–328 (2010)
11. Y. Luchko, Operational rules for a mixed operator of the Erdélyi-Kober type. Fract. Calc. Appl. Analysis **7(3)**, 339–364 (2004)
12. Y. Luchko, J.J. Trujillo, Caputo-type modification of the Erdélyi-Kober fractional derivative. Fract. Calc. Appl. Anal. **10(3)**, 250–267 (2007)
13. F. Mainardi, Yu. Luchko, G. Pagnini, The fundamental solution of the space-time fractional diffusion equation. Fract. Calc. Appl. Anal. **4**, 153–192 (2001)
14. A.M. Mathai, Pathway to matrix-variate gamma and normal densities. Linear Algebra Appl. **396**, 317–328 (2005)
15. A.M. Mathai, Generalized Krätzel integral and associated statistical densities. Int. J. Math. Anal. **6(51)**, 2501–2510 (2012)
16. A.M. Mathai, Fractional integral operators in the complex matrix-variate case. Linear Algebra Appl. **439**, 2901–2913 (2013)
17. A.M. Mathai, Fractional integral operators involving many matrix variables. Linear Algebra Appl. **446**, 196–215 (2014)
18. A.M. Mathai, H.J. Haubold, *Special Functions for Applied Scientists* (Springer, New York, 2008)
19. A.M. Mathai, H.J. Haubold, Stochastic processes via pathway model. Entropy **17**, 2642–2654 (2015)
20. A.M. Mathai, S.B. Provost, T. Hayakawa, *Bilinear Forms and Zonal Polynomials*. Lecture Notes in Statistics (Springer, New York, 1995)
21. A.M. Mathai, R.K. Saxena, H.J. Haubold, *The H-function: Theory and Applications* (Springer, New York, 2010)
22. R. Metzler, W.G. Glöckle, T.F. Nonnenmacher, Fractional model equation for anomalous diffusion. Physica A **211**, 13–24 (1994)
23. K.S. Miller, B. Ross, *An Introduction to the Fractional Calculus and Fractional Differential Equations* (Wiley, New York, 1993)
24. K. Nishimoto (1984/1987/1989/1991/1996) *Fractional Calculus*, vols. 1–5 (Descartes Press, Koriyama, 1996)

25. K.B. Oldham, J. Spanier, *The Fractional Calculus: Theory and Applications of Differentiation and Integration to Arbitrary Order* (Academic, New York, 1974)
26. G. Pagnini, Erdélyi-Kober fractional diffusion. Fract. Calc. Appl. Anal. **15(1)**, 117–127 (2012)
27. I. Podlubny, *Fractional Differential Equations* (Academic, San Diego, 1999)
28. L. Plociniczak, Approximation of the Erdélyi-Kober operator with application to the time-fractional porous medium equation. SIAM J. Appl. Math. **74(4)**, 129–1237 (2014)
29. M. Saigo, A.A. Kilbas, Generalized fractional calculus of the H-function. Fukuoka Univ. Sci. Rep. **29**, 31–45 (1999)
30. I.N. Sneddon, The use in mathematical physics of Erdélyi-Kober operators and of some of their generalizations, in *Fractional Calculus and Its Applications*, ed. by B. Ross (Springer, New York, 1975)
31. H.M. Srivastava, R.K. Saxena, Operators of fractional integration and their applications. Appl. Math. Comput. **118**, 1–52 (2001)

Chapter 3
Erdélyi-Kober Fractional Integrals in the Real Matrix-Variante Case

General notations on matrices, determinants, traces etc. are given in the introduction to Chap. 2 and hence they will not be repeated here. Before starting the discussion, we will need some Jacobians of matrix transformations here. For results on Jacobians, see Mathai [3]. For the real matrix-variate case, the determinant of X will be denoted by either $\det(X)$ or by $|X|$. When complex matrices are involved we will use the notation $\det(X)$ for determinant because we would like to reserve the notation $|(\cdot)|$ for the absolute value of $(\cdot)$. In this case the absolute value of the determinant of $\tilde{X}$ will be written as $|\det(\tilde{X})|$, denoting a matrix X in the complex domain as $\tilde{X}$. All matrices appearing in this chapter are $p \times p$ real positive definite unless stated otherwise. Some Jacobians of matrix transformations will be stated here as lemmas without proofs. For proofs and other details, see Mathai [3].

Lemma 3.1 *Let A be $m \times m$ and B be $n \times n$ nonsingular constant matrices. Let $X = (x_{ij})$ and $Y = (y_{ij})$ be $m \times n$ matrices of mn distinct real variables as elements. Then*

$$Y = AXB, |A| \neq 0, |B| \neq 0, \Rightarrow \mathrm{d}Y = |A|^n |B|^m \mathrm{d}X. \tag{3.1}$$

Lemma 3.2 *Let $X = X'$ and $Y = Y'$ be real symmetric $p \times p$ matrices with $p(p+1)/2$ distinct elements as real scalar variables, where a prime denotes the transpose. Let A be a $p \times p$ nonsingular constant matrix. Then*

$$Y = AXA', |A| \neq 0 \Rightarrow \mathrm{d}Y = |A|^{p+1} \mathrm{d}X. \tag{3.2}$$

Lemma 3.3 *Let X be $p \times p$ and nonsingular. Then*

$$Y = X^{-1} \Rightarrow \mathrm{d}Y = \begin{cases} |X|^{-2p} \mathrm{d}X \textit{ for a general } X \\ |X|^{-(p+1)} \mathrm{d}X \textit{ for } X = X'. \end{cases} \tag{3.3}$$

A. M. Mathai, H. J. Haubold, *Erdélyi–Kober Fractional Calculus*, SpringerBriefs in Mathematical Physics 31, https://doi.org/10.1007/978-981-13-1159-8_3

We will denote the unique positive definite square root of a real positive definite matrix A by $A^{\frac{1}{2}}$. The following standard property will be used very often in this paper. For $p \times p$ nonsingular matrices A and B

$$|I \pm AB| = |I \pm BA| = |A|\,|A^{-1} \pm B| = |B|\,|B^{-1} \pm A| \text{ (when nonsingular)}$$
$$|I \pm AB| = |I \pm A^{\frac{1}{2}}BA^{\frac{1}{2}}| = |I \pm B^{\frac{1}{2}}AB^{\frac{1}{2}}| \text{ (when positive definite).} \tag{3.4}$$

Lemma 3.4 *Let the $p \times p$ matrix X be real positive definite. Let $T = (t_{ij})$ be a lower triangular matrix with t_{ij}'s , $i > j$ be distinct real variables, $t_{ij} = 0, i < j$ and the diagonal elements be positive, $t_{jj} > 0$, $j = 1, \ldots, p$. Then we can show that the transformation $X = TT'$ is unique. Then*

$$X = TT' \Rightarrow \mathrm{d}X = 2^p\{\prod_{j=1}^{p} t_{jj}^{p+1-j}\}\mathrm{d}T. \tag{3.4}$$

By using Lemma 3.4 we can evaluate a real matrix-variate gamma integral, denoted by $\Gamma_p(\alpha)$ where

$$\Gamma_p(\alpha) = \int_{X>O} |X|^{\alpha-\frac{p+1}{2}} \mathrm{e}^{-\mathrm{tr}(X)}\mathrm{d}X. \tag{3.5}$$

Apply Lemma 3.4 to the right side of (3.5). Then the integral splits into p integrals on t_{jj}'s and $p(p-1)/2$ integrals on t_{ij}'s, $i > j$. The t_{jj}-integral gives $\Gamma(\alpha - \frac{j-1}{2})$, $\Re(\alpha) > \frac{j-1}{2}$, $j = 1, \ldots, p$ which gives the final condition as $\Re(\alpha) > \frac{p-1}{2}$. Each of the t_{ij}-integral for $i > j$ gives $\sqrt{\pi}$ and there are $p(p-1)/2$ of them, thus giving the final factor $\pi^{\frac{p(p-1)}{4}}$. Hence

$$\begin{aligned}\Gamma_p(\alpha) &= \int_{X>O} |X|^{\alpha-\frac{p+1}{2}} \mathrm{e}^{-\mathrm{tr}(X)}\mathrm{d}X \\ &= \pi^{\frac{p(p-1)}{4}} \Gamma(\alpha)\Gamma(\alpha - \frac{1}{2}) \ldots \Gamma(\alpha - \frac{p-1}{2}), \Re(\alpha) > \frac{p-1}{2}.\end{aligned} \tag{3.6}$$

Lemma 3.5 *Let Y be $p \times n$, $n \geq p$ and be of rank p or let Y be a full rank matrix of np distinct real scalar variables as elements. Let $S = YY'$. Then $S > O$ (positive definite). Then writing $Y = TU$ where T is a $p \times p$ lower triangular matrix with positive diagonal elements and U is $p \times n$ semi-orthonormal matrix so that $S = YY' = TUU'T' = TT'$. Then integrating out the differential elements coming from the semi-orthonormal matrix U or integrating out over the Stiefel manifold and then substituting for $\mathrm{d}T$, we can connect the differential elements $\mathrm{d}S$ and $\mathrm{d}Y$. The result is the following:*

$$\mathrm{d}Y = \frac{\pi^{\frac{np}{2}}}{\Gamma_p(\frac{n}{2})}|S|^{\frac{n}{2}-\frac{p+1}{2}}\mathrm{d}S$$

where $\Gamma_p(\frac{n}{2})$ is defined in (3.6).

This Lemma 3.5 is very important in the theory of functions of matrix argument. This lemma enables us to extend the results from positive definite matrices to rectangular matrices of full rank. There are many applications of this result in different disciplines, some of which may be seen from Mathai [4, 5], Mathai and Princy [13, 14].

We need real matrix-variate type-1 and type-2 beta functions, beta integrals and beta densities in our discussion later on. Hence these will be given here. The real matrix-variate type-1 beta density for the $p \times p$ real positive definite matrix X_1, with parameters α and β and denoted by $f_1(X_1)$, is defined as follows:

$$f_1(X_1) = \frac{\Gamma_p(\alpha+\beta)}{\Gamma_p(\alpha)\Gamma_p(\beta)}|X_1|^{\alpha-\frac{p+1}{2}}|I-X_1|^{\beta-\frac{p+1}{2}},\ O < X_1 < I \text{ (type-1)} \tag{3.7}$$

$$f_2(Y_1) = \frac{\Gamma_p(\alpha+\beta)}{\Gamma_p(\alpha)\Gamma_p(\beta)}|Y_1|^{\beta-\frac{p+1}{2}}|I-Y_1|^{\alpha-\frac{p+1}{2}}\mathrm{d}Y_1,\ O < Y_1 < I \text{ (type-1)}$$

for $\Re(\alpha) > \frac{p-1}{2}$, $\Re(\beta) > \frac{p-1}{2}$, and $f(X_1) = 0$, $f_2(Y_1) = 0$ elsewhere. Type-1 and Type-2 beta integrals and beta functions are defined and denoted as follows for $\Re(\alpha) > \frac{p-1}{2}$, $\Re(\beta) > \frac{p-1}{2}$, where $B_p(\alpha, \beta)$ is the real matrix-variate beta function which is defined in terms of gamma functions as given below:

$$\begin{aligned}
B_p(\alpha,\beta) &= \frac{\Gamma_p(\alpha)\Gamma_p(\beta)}{\Gamma_p(\alpha+\beta)} \\
&= \int_{O<X<I}|X|^{\alpha-\frac{p+1}{2}}|I-X|^{\beta-\frac{p+1}{2}}\mathrm{d}X \text{ (type-1)} \\
&= \int_{O<Y<I}|Y|^{\beta-\frac{p+1}{2}}|I-Y|^{\alpha-\frac{p+1}{2}}\mathrm{d}Y \text{ (type-1)} \\
&= \int_{U>O}|U|^{\alpha-\frac{p+1}{2}}|I+U|^{-(\alpha+\beta)}\mathrm{d}U \text{ (type-2)} \\
&= \int_{V>O}|V|^{\beta-\frac{p+1}{2}}|I+V|^{-(\alpha+\beta)}\mathrm{d}V \text{ (type-2)}
\end{aligned} \tag{3.8}$$

3.1 Explicit Evaluations of Matrix-Variate Gamma and Beta Integrals

The integrals in (3.6) and (3.8) look complicated as multivariate integrals. Is it possible to evaluate these integrals by integrating out over the elements of the matrices or as multiple integrals or as multivariate real scalar variable integrals? If evaluated over the individual real scalar variables, is it going to give the same results in terms of gamma products such as the ones in (3.6)? We will examine this aspect in this section.

Matrix transformations in terms of triangular matrices is the easiest way of evaluating matrix-variate gamma and beta integrals in the real cases. Here we give several procedures of explicit evaluation of gamma and beta integrals in the general real situations. The procedure also reveals the structure of these matrix-variate integrals. Apart from the evaluation of matrix-variate gamma and beta integrals, the procedure can also be applied to evaluate such integrals explicitly in similar situations. Various methods described here will be useful to those who are working on integrals involving real-valued scalar functions of matrix argument in general and gamma and beta integrals in particular.

First we consider matrix-variate gamma integrals in the real case, then we look at matrix-variate type-1 beta integrals in the real case. The procedure is parallel in the matrix-variate type-2 beta integrals.

3.1.1 Explicit Evaluation of Real Matrix-Variate Gamma Integral

Matrix-variate gamma integral is a very popular integral in many areas. A particular case is the most popular Wishart density in multivariate statistical analysis. Let X be a $p \times p$ real symmetric and positive definite matrix of mathematical or random variables. Consider the real-valued function of matrix argument

$$f(X) = C\ |X|^{\alpha-\frac{p+1}{2}} \mathrm{e}^{-\mathrm{tr}(BX)} \qquad (a)$$

where C is a constant. When X is real and positive definite, $X > O$, then $f(X)$ in (a) represents a real matrix-variate gamma density when $C = \frac{|B|^{\alpha}}{\Gamma_p(\alpha)}$ where $B > O$ is a constant matrix. When B is of the form $B = \frac{1}{2}V^{-1},\ V = V' > O$, where a prime denotes the transpose, then $f(X)$ in (a) is the Wishart density in multivariate statistical analysis, which is the central density in the area, see for example, Anderson [1], Kshirsagar [2], Srivastava and Khatri [17]. The real matrix-variate gamma integral is given in (3.6). The integral in (3.6) is evaluated there by using the standard technique of writing $X = TT'$ where T is a lower or upper triangular matrix with positive diagonal elements and evaluating the integrals over t_{ij}'s as discussed after (3.6).

When Wishart density is derived, starting from samples from a Gaussian population, the basic technique is the triangularization process. Can we evaluate the integral on the right of (a) or (3.6) by using conventional methods, or by direct evaluation? We will look into this problem by using the technique of partitioned matrices, see also Mathai [8, 9]. Let us partition

$$X = \begin{bmatrix} X_{11} & X_{12} \\ X_{21} & X_{22} \end{bmatrix}$$

where let $X_{22} = x_{pp}$ so that $X_{21} = (x_{p1}, \ldots, x_{pp-1})$, $X_{12} = X_{21}'$. Then

$$|X|^{\alpha-\frac{p+1}{2}} = |X_{11}|^{\alpha-\frac{p+1}{2}}[x_{pp} - X_{21}X_{11}^{-1}X_{12}]^{\alpha-\frac{p+1}{2}}$$

by using partitioned matrix and determinant. Note that when X is positive definite, that is, $X > O$, then $X_{11} > O, x_{pp} > 0$ and the quadratic form $X_{21}X_{11}^{-1}X_{12} > 0$. Note that

$$[x_{pp} - X_{21}X_{11}^{-1}X_{12}]^{\alpha-\frac{p+1}{2}} = x_{pp}^{\alpha-\frac{p+1}{2}}[1 - x_{pp}^{-\frac{1}{2}}X_{21}X_{11}^{-\frac{1}{2}}X_{11}^{-\frac{1}{2}}X_{12}x_{pp}^{-\frac{1}{2}}]^{\alpha-\frac{p+1}{2}}.$$

Let $Y = x_{pp}^{-\frac{1}{2}}X_{21}X_{11}^{-\frac{1}{2}}$ then $\mathrm{d}Y = x_{pp}^{-\frac{p-1}{2}}|X_{11}|^{-\frac{1}{2}}\mathrm{d}X_{21}$ for fixed X_{11}, x_{pp}, see Mathai (1997, Theorem 1.18.) [3] or Lemma 3.1. The integral over x_{pp} gives

$$\int_0^\infty x_{pp}^{\alpha+\frac{p-1}{2}-\frac{p+1}{2}}\mathrm{e}^{-x_{pp}}\mathrm{d}x_{pp} = \Gamma(\alpha),\ \Re(\alpha) > 0.$$

Let $u = YY'$. Then from Lemma 3.5 or from Theorem 2.16 and Remark 2.13 of Mathai [3] and after integrating out over the Stiefel manifold we have

$$\mathrm{d}Y = \frac{\pi^{\frac{p-1}{2}}}{\Gamma(\frac{p-1}{2})}u^{\frac{p-1}{2}-1}\mathrm{d}u.$$

(Note that n in Theorem 2.16 of Mathai [3] corresponds to $p-1$ and p corresponds to 1). Then the integral over u gives

$$\int_0^1 u^{\frac{p-1}{2}-1}(1-u)^{\alpha-\frac{p+1}{2}}\mathrm{d}u = \frac{\Gamma(\frac{p-1}{2})\Gamma(\alpha-\frac{p-1}{2})}{\Gamma(\alpha)},\ \Re(\alpha) > \frac{p-1}{2},$$

from a real scalar variable type-1 beta integral. Now, collecting all the factors, we have

$$|X_{11}|^{\alpha+\frac{1}{2}-\frac{p+1}{2}}\Gamma(\alpha)\frac{\pi^{\frac{p-1}{2}}}{\Gamma(\frac{p-1}{2})}\frac{\Gamma(\frac{p-1}{2})\Gamma(\alpha-\frac{p-1}{2})}{\Gamma(\alpha)} = |X_{11}^{(1)}|^{\alpha+\frac{1}{2}-\frac{p+1}{2}}\pi^{\frac{p-1}{2}}\Gamma(\alpha-\frac{p-1}{2})$$

for $\Re(\alpha) > \frac{p-1}{2}$. Note that $|X_{11}^{(1)}|$ is $(p-1)\times(p-1)$ and $|X_{11}|$ after the completion of the first part of the operations is denoted by $|X_{11}^{(1)}|$, and the exponent is changed to $\alpha+\frac{1}{2}-\frac{p+1}{2}$. Now repeat the process by separating $x_{p-1,p-1}$, that is by writing

$$X_{11}^{(1)} = \begin{bmatrix} X_{11}^{(2)} & X_{12}^{(2)} \\ X_{21}^{(2)} & x_{p-1,p-1} \end{bmatrix}.$$

Here $X_{11}^{(2)}$ is of order $(p-2)\times(p-2)$ and $X_{21}^{(2)}$ is of order $1\times(p-2)$. As before, let $u=YY', Y=x_{p-1,p-1}^{-\frac{1}{2}}X_{21}^{(2)}[X_{11}^{(2)}]^{-\frac{1}{2}}$. Then $\mathrm{d}Y=x_{p-1,p-1}^{-\frac{p-2}{2}}|X_{11}^{(2)}|^{-\frac{1}{2}}\mathrm{d}X_{21}^{(2)}$. Integral over the Stiefel manifold gives $\frac{\pi^{\frac{p-2}{2}}}{\Gamma(\frac{p-2}{2})}u^{\frac{p-2}{2}-1}\mathrm{d}u$ and the factor containing $(1-u)$ is $(1-u)^{\alpha+\frac{1}{2}-\frac{p+1}{2}}$ and the integral over u gives

$$\int_0^1 u^{\frac{p-2}{2}-1}(1-u)^{\alpha+\frac{1}{2}-\frac{p+1}{2}}\mathrm{d}u=\frac{\Gamma(\frac{p-2}{2})\Gamma(\alpha-\frac{p-2}{2})}{\Gamma(\alpha)}.$$

Intgral over $v=x_{p-1,p-1}$ gives

$$\int_0^1 v^{\alpha+\frac{1}{2}+\frac{p-2}{2}-\frac{p+1}{2}}\mathrm{e}^{-v}\mathrm{d}v=\Gamma(\alpha),\ \Re(\alpha)>0.$$

Taking all product we have

$$|X_{11}^{(2)}|^{\alpha+1-\frac{p+1}{2}}\pi^{\frac{p-2}{2}}\Gamma(\alpha-\frac{p-2}{2}),\ \Re(\alpha)>\frac{p-2}{2}.$$

Successive evaluations by using the same procedure gives the exponent of π as $\frac{p-1}{2}+\frac{p-2}{2}+\ldots+\frac{1}{2}=\frac{p(p-1)}{4}$ and the gamma product is $\Gamma(\alpha-\frac{p-1}{2})\Gamma(\alpha-\frac{p-2}{2})\ldots\Gamma(\alpha)$ and the final result is $\Gamma_p(\alpha)$. Hence the result is verified.

3.1.2 Evaluation of Matrix-Variate Type-1 Beta Integral in the Real Case

The real matrix-variate type-1 beta density is available from (3.7) and type-1 beta integrals from (3.8). For evaluating real matrix-variate gamma integral an easy method is to make the transformation $X=TT'$ where T is a lower triangular matrix with positive diagonal elements. Even if this transformation is applied here in the case of beta integrals, the integral does not simplify due to the presence of the factor $|I-X|^{\beta-\frac{p+1}{2}}$. Hence we will try to evaluate the integral by using a partitioning of the matrices and then integrating step by step. Let $X=(x_{ij})$ be a $p\times p$ matrix. Let us separate x_{pp}. This can be done by partitioning $|X|$ and $|I-X|$. That is, let

$$X=\begin{bmatrix}X_{11} & X_{12}\\ X_{21} & X_{22}\end{bmatrix}$$

where X_{11} is the $(p-1)\times(p-1)$ leading sub-matrix, X_{21} is $1\times(p-1)$, $X_{22}=x_{pp}$ and $X_{12}=X_{21}'$. Then $|X|=|X_{11}|[x_{pp}-X_{21}X_{11}^{-1}X_{12}]$ and

$$|X|^{\alpha-\frac{p+1}{2}} = |X_{11}|^{\alpha-\frac{p+1}{2}}[x_{pp} - X_{21}X_{11}^{-1}X_{12}]^{\alpha-\frac{p+1}{2}} \qquad (i)$$

$$|I - X|^{\beta-\frac{p+1}{2}} = |I - X_{11}|^{\beta-\frac{p+1}{2}}[(1 - x_{pp}) - X_{21}(I - X_{11})^{-1}X_{12}]^{\beta-\frac{p+1}{2}} \qquad (ii)$$

From (i) we have $x_{pp} > X_{21}X_{11}^{-1}X_{12}$ and from (ii) we have $x_{pp} < 1 - X_{21}(I - X_{11})^{-1}X_{12}$. That is, $X_{21}X_{11}^{-1}X_{12} < x_{pp} < 1 - X_{21}(I - X_{11})^{-1}X_{12}$. Let $y = x_{pp} - X_{21}X_{11}^{-1}X_{12} \Rightarrow \mathrm{d}y = \mathrm{d}x_{pp}$ for fixed X_{21}, X_{11}. Also, $0 < y < b$ where

$$\begin{aligned} b &= 1 - X_{21}X_{11}^{-1}X_{12} - X_{21}(I - X_{11})^{-1}X_{12} \\ &= 1 - X_{21}X_{11}^{-\frac{1}{2}}(I - X_{11})^{-\frac{1}{2}}(I - X_{11})^{-\frac{1}{2}}X_{11}^{-\frac{1}{2}}X_{12} \\ &= 1 - WW', \; W = X_{21}X_{11}^{-\frac{1}{2}}(I - X_{11})^{-\frac{1}{2}}. \end{aligned}$$

The second factor on the right in (ii) becomes

$$[b - y]^{\beta-\frac{p+1}{2}} = b^{\beta-\frac{p+1}{2}}[1 - \frac{y}{b}]^{\beta-\frac{p+1}{2}}.$$

Put $u = \frac{y}{b}$ for fixed b. Then the factors containing u and b become $b^{\alpha+\beta-(p+1)+1}u^{\alpha-\frac{p+1}{2}}(1 - u)^{\beta-\frac{p+1}{2}}$. Integral over u gives

$$\int_0^1 u^{\alpha-\frac{p+1}{2}}(1 - u)^{\beta-\frac{p+1}{2}}\mathrm{d}u = \frac{\Gamma(\alpha - \frac{p-1}{2})\Gamma(\beta - \frac{p-1}{2})}{\Gamma(\alpha + \beta - (p - 1))},$$

for $\Re(\alpha) > \frac{p-1}{2}$, $\Re(\beta) > \frac{p-1}{2}$. Let $W = X_{21}X_{11}^{-\frac{1}{2}}(I - X_{11})^{-\frac{1}{2}}$ for fixed X_{11}. Then $\mathrm{d}X_{21} = |X_{11}|^{\frac{1}{2}}|I - X_{11}|^{\frac{1}{2}}\mathrm{d}W$ from Lemma 2.1 or from Theorem 1.18 of Mathai [3], where X_{11} is $(p - 1) \times (p - 1)$. Put $v = WW'$ and integrate out over the Stiefel manifold by using Lemma 2.5 or Theorem 2.16 and Remark 2.13 of Mathai [3]. Then we have

$$\mathrm{d}W = \frac{\pi^{\frac{p-1}{2}}}{\Gamma(\frac{p-1}{2})}v^{\frac{p-1}{2}-1}\mathrm{d}v.$$

Now the integral over b becomes

$$\int b^{\alpha+\beta-p}\mathrm{d}X_{21} = \int_0^1 v^{\frac{p-1}{2}-1}(1-v)^{\alpha+\beta-p}\mathrm{d}v = \frac{\Gamma(\frac{p-1}{2})\Gamma(\alpha + \beta - (p - 1))}{\Gamma(\alpha + \beta - \frac{p-1}{2})}, \Re(\alpha + \beta) > p - 1.$$

Now, multiplying all the factors together we have

$$|X_{11}^{(1)}|^{\alpha+\frac{1}{2}-\frac{p+1}{2}}|I - X_{11}^{(1)}|^{\beta+\frac{1}{2}-\frac{p+1}{2}}\pi^{\frac{p-1}{2}}\frac{\Gamma(\alpha - \frac{p-1}{2})\Gamma(\beta - \frac{p-1}{2})}{\Gamma(\alpha + \beta - \frac{p-1}{2})}$$

for $\Re(\alpha) > \frac{p-1}{2}$, $\Re(\beta) > \frac{p-1}{2}$. Here $X_{11}^{(1)}$ indicates the $(p-1) \times (p-1)$ leading sub-matrix at the end of the first set of operations. At the end of the second set of operations we will denote the $(p-2) \times (p-2)$ leading sub-matrix by $X_{11}^{(2)}$, and so on. The second step of operations starts by separating $x_{p-1,p-1}$ and writing

$$|X_{11}^{(1)}| = |X_{11}^{(2)}|[x_{p-1,p-1} - X_{21}^{(2)}[X_{11}^{(2)}]^{-1}X_{12}^{(2)}]$$

where $X_{21}^{(2)}$ is $1 \times (p-2)$. Now, proceed as in the first sequence of steps to obtain the final factors of the following form:

$$|X_{11}^{(2)}|^{\alpha+1-\frac{p+1}{2}}|I - X_{11}^{(2)}|^{\beta+1-\frac{p+1}{2}}\pi^{\frac{p-2}{2}}\frac{\Gamma(\alpha - \frac{p-2}{2})\Gamma(\beta - \frac{p-2}{2})}{\Gamma(\alpha+\beta-\frac{p-2}{2})}$$

for $\Re(\alpha) > \frac{p-2}{2}$, $\Re(\beta) > \frac{p-2}{2}$. Proceeding like this the exponent of π at the end will be of the form

$$\frac{p-1}{2} + \frac{p-2}{2} + \ldots + \frac{1}{2} = \frac{p(p-1)}{4}.$$

The gamma product will be of the form

$$\frac{\Gamma(\alpha - \frac{p-1}{2})\Gamma(\alpha - \frac{p-2}{2})\ldots\Gamma(\alpha)\Gamma(\beta - \frac{p-1}{2})\ldots\Gamma(\beta)}{\Gamma(\alpha+\beta-\frac{p-1}{2})\ldots\Gamma(\alpha+\beta)}.$$

These gamma products, together with $\pi^{\frac{p(p-1)}{4}}$ can be written as $\frac{\Gamma_p(\alpha)\Gamma_p(\beta)}{\Gamma_p(\alpha+\beta)} = B_p(\alpha, \beta)$ and hence the result. Thus, it is possible to evaluate the type-1 real matrix-variate beta integral directly to obtain the beta function in the real matrix-variate case.

A similar procedure can yield the real matrix-variate beta function from the type-2 real matrix-variate beta integral of the form

$$\int_{X>O}|X|^{\alpha-\frac{p+1}{2}}|I+X|^{-(\alpha+\beta)}\mathrm{d}X$$

for $X = X' > O$ and $p \times p$, $\Re(\alpha) > \frac{p-1}{2}$, $\Re(\beta) > \frac{p-1}{2}$. The procedure for the evaluation will be parallel.

3.1.3 General Partitions

In Sects. 3.1.1 and 3.1.2, we have considered integrating one variable at a time by suitably partitioning the matrices. Is it possible to have a general partitioning and integrate a block of variables at a time, rather than integrating out individual variables? Let us consider the real matrix-variate gamma integral first. Let

$$X = \begin{bmatrix} X_{11} & X_{12} \\ X_{21} & X_{22} \end{bmatrix}, \ X_{11} \text{ is } p_1 \times p_1 \text{ and } X_{22} \text{ is } p_2 \times p_2$$

so that X_{12} is $p_1 \times p_2$ and $X_{21} = X_{12}'$ and $p_1 + p_2 = p$. Without loss of generality, let us assume that $p_1 \geq p_2$. Then the determinant can be partitioned as follows:

$$\begin{aligned} |X|^{\alpha-\frac{p+1}{2}} &= |X_{11}|^{\alpha-\frac{p+1}{2}}|X_{22} - X_{21}X_{11}^{-1}X_{12}|^{\alpha-\frac{p+1}{2}} \\ &= |X_{11}|^{\alpha-\frac{p+1}{2}}|X_{22}|^{\alpha-\frac{p+1}{2}}|I - X_{22}^{-\frac{1}{2}}X_{21}X_{11}^{-1}X_{12}X_{22}^{-\frac{1}{2}}|^{\alpha-\frac{p+1}{2}}. \end{aligned}$$

Put

$$Y = X_{22}^{-\frac{1}{2}}X_{21}X_{11}^{-\frac{1}{2}} \Rightarrow \mathrm{d}Y = |X_{22}|^{-\frac{p_1}{2}}|X_{11}|^{-\frac{p_2}{2}}\mathrm{d}X_{21}$$

for fixed X_{11} and X_{22}.

$$|X|^{\alpha-\frac{p+1}{2}} = |X_{11}|^{\alpha+\frac{p_2}{2}-\frac{p+1}{2}}|X_{22}|^{\alpha+\frac{p_1}{2}-\frac{p+1}{2}}|I - YY'|^{\alpha-\frac{p+1}{2}}.$$

The Jacobian above is available from Lemma 3.1 or from Theorem 1.18 of Mahai [3]. Let $S = YY'$. Then integrating out over the Stiefel manifold we have

$$\mathrm{d}Y = \frac{\pi^{\frac{p_1p_2}{2}}}{\Gamma_{p_2}(\frac{p_1}{2})}|S|^{\frac{p_1}{2}-\frac{p_2+1}{2}}\mathrm{d}S,$$

see Lemma 3.5 or Theorem 2.16 and Remark 2.13 of Mathai [3]. Now, integral over S gives

$$\int_{O<S<I}|S|^{\frac{p_1}{2}-\frac{p_2+1}{2}}|I - S|^{\alpha-\frac{p_1}{2}-\frac{p_2+1}{2}}\mathrm{d}S = \frac{\Gamma_{p_2}(\frac{p_1}{2})\Gamma_{p_2}(\alpha-\frac{p_1}{2})}{\Gamma_{p_2}(\alpha)},$$

for $\Re(\alpha) > \frac{p_1-1}{2}$. Collecting all the factors, we have

$$|X_{11}|^{\alpha-\frac{p_1+1}{2}}|X_{22}|^{\alpha-\frac{p_2+1}{2}}\pi^{\frac{p_1p_2}{2}}\frac{\Gamma_{p_2}(\alpha-\frac{p_1}{2})}{\Gamma_{p_2}(\alpha)}.$$

From here, one can also observe that the original determinant splits into functions of X_{11} and X_{22}. This also shows that if we are considering a real matrix-variate gamma density then the diagonal blocks X_{11} and X_{22} are statistically independently distributed, where X_{11} will have a p_1-variate gamma distribution and X_{22} has a p_2-variate gamma distribution. Observe that $\mathrm{tr}(X) = \mathrm{tr}(X_{11}) + \mathrm{tr}(X_{22})$ and hence the integral over X_{22} gives $\Gamma_{p_2}(\alpha)$ and the integral over X_{11} gives $\Gamma_{p_1}(\alpha)$. Hence the total integral is available as

$$\Gamma_{p_1}(\alpha)\Gamma_{p_2}(\alpha)\pi^{\frac{p_1p_2}{2}}\frac{\Gamma_{p_2}(\alpha-\frac{p_1}{2})}{\Gamma_{p_2}(\alpha)}=\Gamma_p(\alpha)$$

since $\pi^{\frac{p_1p_2}{2}}\Gamma_{p_1}(\alpha)\Gamma_{p_2}(\alpha-\frac{p_1}{2})=\Gamma_p(\alpha)$.

Hence it is seen that instead of integrating out variables one at a time we could have also integrated out blocks of variables at a time and could have verified the result. Similar procedure works for real matrix-variate type-1 and type-2 beta also.

3.1.4 A Method of Avoiding Integration Over the Stiefel Manifold

The general method of partitioning described above involves the integration over the Stiefel manifold as an intermediate step of substituting for the differential element of a lower triangular matrix. Another method will be considered here, which will avoid integration over Stiefel manifold. Let us consider the real matrix-variate gamma case first. Again, we start with the decomposition

$$|X|^{\alpha-\frac{p+1}{2}}=|X_{11}|^{\alpha-\frac{p+1}{2}}|X_{22}-X_{21}X_{11}^{-1}X_{12}|^{\alpha-\frac{p+1}{2}}. \qquad (iii)$$

Instead of integrating out X_{21} or X_{12} let us integrate out X_{22}. Let X_{11} be $p_1\times p_1$ and X_{22} be $p_2\times p_2$ with $p_1+p_2=p$. In the above partitioning we require that X_{11} be nonsingular. But when X is positive definite, both X_{11} and X_{22} will be positive definite, thereby nonsingular also. From the second factor in (iii), $X_{22}>X_{21}X_{11}^{-1}X_{12}$ from $X_{22}-X_{21}X_{11}^{-1}X_{12}$ being positive definite. We will try to integrate out X_{22} first. Let $U=X_{22}-X_{21}X_{11}^{-1}X_{12}$ so that $\mathrm{d}U=\mathrm{d}X_{22}$ for fixed X_{11} and X_{12}. Since $\mathrm{tr}(X)=\mathrm{tr}(X_{11})+\mathrm{tr}(X_{22})$ we have

$$\mathrm{e}^{-\mathrm{tr}(X_{22})}=\mathrm{e}^{-\mathrm{tr}(U)-\mathrm{tr}(X_{21}X_{11}^{-1}X_{12})}.$$

Integrating out U we have

$$\int_{U>O}|U|^{\alpha-\frac{p+1}{2}}\mathrm{e}^{-\mathrm{tr}(U)}\mathrm{d}U=\Gamma_{p_2}(\alpha-\frac{p_1}{2}),\ \Re(\alpha)>\frac{p-1}{2}$$

since $\alpha-\frac{p+1}{2}=\alpha-\frac{p_1}{2}-\frac{p_2+1}{2}$. Let

$$Y=X_{21}X_{11}^{-\frac{1}{2}}\Rightarrow \mathrm{d}Y=|X_{11}|^{-\frac{p_2}{2}}\mathrm{d}X_{21}$$

for fixed X_{11}. Then

$$\int_{X_{21}} \mathrm{e}^{-\mathrm{tr}(X_{21}X_{11}^{-1}X_{12})}\mathrm{d}X_{21} = |X_{11}|^{\frac{p_2}{2}} \int_Y \mathrm{e}^{-\mathrm{tr}(YY')}\mathrm{d}Y.$$

But $\mathrm{tr}(YY')$ is the sum of squares of the p_1p_2 elements in Y and each integral is of the form $\int_{-\infty}^{\infty} \mathrm{e}^{-z^2}\mathrm{d}z = \sqrt{\pi}$. Hence

$$\int_Y \mathrm{e}^{-\mathrm{tr}(YY')}\mathrm{d}Y = \pi^{\frac{p_1p_2}{2}}.$$

Now we can integrate out X_{11}.

$$\int_{X_{11}>O} |X_{11}|^{\alpha+\frac{p_2}{2}-\frac{p+1}{2}} \mathrm{e}^{-\mathrm{tr}(X_{11})}\mathrm{d}X_{11} = \int_{X_{11}>O} |X_{11}|^{\alpha-\frac{p_1+1}{2}} \mathrm{e}^{-\mathrm{tr}(X_{11})}\mathrm{d}X_{11} = \Gamma_{p_1}(\alpha).$$

Hence we have the following factors:

$$\pi^{\frac{p_1p_2}{2}}\Gamma_{p_2}(\alpha-\frac{p_1}{2})\Gamma_{p_1}(\alpha) = \Gamma_p(\alpha)$$

since

$$\frac{p_1(p_1-1)}{4} + \frac{p_2(p_2-1)}{4} + \frac{p_1p_2}{2} = \frac{p(p-1)}{4}, \ p = p_1 + p_2$$

and

$$\begin{aligned}\Gamma_{p_1}(\alpha)\Gamma_{p_2}(\alpha-\frac{p_1}{2}) &= \Gamma(\alpha)\Gamma(\alpha-\frac{1}{2})\ldots\Gamma(\alpha-\frac{p_1-1}{2})\Gamma_{p_2}(\alpha-\frac{p_1}{2}) \\ &= \Gamma(\alpha)\ldots\Gamma(\alpha-\frac{p_1+p_2-1}{2}).\end{aligned}$$

Hence the result. In this procedure we did not have to go through integration over the Stiefel manifold and we did not have to assume that $p_1 \geq p_2$. We could have integrated out X_{11} first if needed. In this case, expand

$$|X|^{\alpha-\frac{p+1}{2}} = |X_{22}|^{\alpha-\frac{p+1}{2}}|X_{11} - X_{12}X_{22}^{-1}X_{21}|^{\alpha-\frac{p+1}{2}}.$$

Then proceed as before by integrating out X_{11} first. Then we end up with

$$\pi^{\frac{p_1p_2}{2}}\Gamma_{p_1}(\alpha-\frac{p_2}{2})\Gamma_{p_2}(\alpha) = \Gamma_p(\alpha), \ p = p_1 + p_2.$$

Note 3.1 If we are considering a real matrix-variate gamma density, such as the Wishart density, then from the above procedure observe that after integrating out X_{22} the only factor containing X_{21} is the exponential function, which has the structure of a matrix-variate Gaussian density. Hence for given X_{11}, X_{21} is matrix-variate Gaussian distributed. Similarly, for given X_{22}, X_{12} is matrix-variate Gaussian distributed. Further, the diagonal blocks X_{11} and X_{22} are independently distributed.

As seen from the above considerations, one can integrate elements one at a time and finally evaluate a matrix-variate gamma or beta integrals or one can integrate blocks of variables at a time and evaluate the matrix-variate gamma and beta integrals. But we will see that the matrix methods adopted in our procedures is the simplest when it comes to real-valued scalar functions with matrix arguments and integrals over them. Some applications of the real matrix-variate integrals are given in Mathai [5–8], Mathai and Haubold [11, 12], Mathai and Princy [13, 14], Thomas and Mathai [16].

3.2 Erdélyi-Kober Fractional Integral Operator of the Second Kind for the Real Matrix-Variate Case

In the real matrix-variate case, it is easier to introduce the second kind fractional integrals compared to the first kind fractional integrals. Hence we will start with the second kind fractional integrals first.

As in the real scalar variable case, we will use the following notation: For Erdélyi-Kober fractional integral or differential operators we will use the letter K. For the first kind or left-sided we use the number 1 and for the second kind or right-sided we use the number 2. The order is denoted by α. When it is a fractional integral the order α will be written as a superscript to K as $-\alpha$ and for fractional derivative the superscript will be written as α or $+\alpha$. The kind number, variable and the additional parameter will be written as subscripts to K. Thus, for example, $K_{1,U,\beta}^{-\alpha}f$ will denote Erdélyi-Kober fractional integral operator of order α and parameter β and of the first kind operating on the arbitrary function f. $K_{2,U,\beta}^{-\alpha}f$ will be Erdélyi-Kober fractional integral of order α, parameter β and of the second kind.

Definition 3.1 We will define and denote Erdélyi-Kober fractional integral operator of the second kind, operating on a real-valued scalar function f of real matrix argument as follows:

$$K_{2,X,\zeta}^{-\alpha}f = \frac{|X|^{\zeta}}{\Gamma_p(\alpha)}\int_{T>X>O}|T-X|^{\alpha-\frac{p+1}{2}}|T|^{-\zeta-\alpha}f(T)\mathrm{d}T, \Re(\alpha) > \frac{p-1}{2}. \tag{3.9}$$

Here $T > X > O$ means that $T > O, X > O, T - X > O$. The definition in (3.9) for $p = 1$ corresponds to Erdélyi-Kober fractional integral of order α and of the second kind and hence we will call (3.9) as the corresponding fractional integral in the real matrix-variate case. Consider two $p \times p$ real positive definite matrix-variate random variables X_1 and X_2, independently distributed, where X_1 has a real matrix-variate type-1 beta density $f_3(X_1)$ with parameters $(\zeta + \frac{p+1}{2}, \alpha)$, that is,

$$f_3(X_1) = \frac{\Gamma_p(\zeta + \alpha + \frac{p+1}{2})}{\Gamma_p(\zeta + \frac{p+1}{2})\Gamma_p(\alpha)} |X_1|^{\zeta} |I - X_1|^{\alpha - \frac{p+1}{2}}, O < X_1 < I$$

for $\Re(\zeta) > -1, \Re(\alpha) > \frac{p-1}{2}$ and $f_3(X_1) = 0$ elsewhere. Let X_2 have the density $f_4(X_2) = f(X_2)$ where f is arbitrary, in the sense $f(X_2) \geq 0$ for all X_2 and $\int_{X_2} f(X_2)\mathrm{d}X_2 = 1$. Then the joint density of X_1 and X_2 is $f_3(X_1)f(X_2)$. Let us consider the transformation $U_2 = X_2^{\frac{1}{2}} X_1 X_2^{\frac{1}{2}}, X_2 = V$ so that $X_2 = V, X_1 = V^{-\frac{1}{2}} U_2 V^{-\frac{1}{2}}$ and the Jacobian is given by $\mathrm{d}X_1 \wedge \mathrm{d}X_2 = |V|^{-\frac{p+1}{2}} \mathrm{d}U_2 \wedge \mathrm{d}V$. If the joint density is denoted by $f(U_2, V)$ then

$$\begin{aligned} f(U_2, V)\mathrm{d}U_2 \wedge \mathrm{d}V &= \frac{\Gamma_p(\zeta + \alpha + \frac{p+1}{2})}{\Gamma_p(\alpha)\Gamma(\zeta + \frac{p+1}{2})} |V^{-\frac{1}{2}} U_2 V^{-\frac{1}{2}}|^{\zeta} \\ &\quad \times |I - V^{-\frac{1}{2}} U_2 V^{-\frac{1}{2}}|^{\alpha - \frac{p+1}{2}} f(V)|V|^{-\frac{p+1}{2}} \mathrm{d}U_2 \wedge \mathrm{d}V. \end{aligned}$$

Here U_2 will be called a symmetric product of the matrices X_1 and X_2. Therefore the marginal density of U_2, denoted by $g_2(U_2)$, is available by integrating out V from $f(U_2, V)$. That is,

$$\begin{aligned} g_2(U_2) &= \int_V f_3(V^{-\frac{1}{2}} U_2 V^{-\frac{1}{2}}) f(V)|V|^{-\frac{p+1}{2}} \mathrm{d}V \\ &= \frac{\Gamma_p(\zeta + \alpha + \frac{p+1}{2})}{\Gamma_p(\zeta + \frac{p+1}{2})} \int_{V > U_2 > O} \frac{1}{\Gamma_p(\alpha)} |V|^{-\frac{p+1}{2}} |U_2|^{\zeta} |V|^{-\zeta} |V|^{-\alpha + \frac{p+1}{2}} |V - U_2|^{\alpha - \frac{p+1}{2}} f(V)\mathrm{d}V \\ g_2(U_2) &= \frac{\Gamma_p(\alpha + \zeta + \frac{p+1}{2})}{\Gamma_p(\zeta + \frac{p+1}{2})} \frac{|U_2|^{\zeta}}{\Gamma_p(\alpha)} \int_{V > U_2 > O} |V|^{-\zeta - \alpha} |V - U_2|^{\alpha - \frac{p+1}{2}} f(V)\mathrm{d}V \\ &= \frac{\Gamma_p(\alpha + \zeta + \frac{p+1}{2})}{\Gamma_p(\zeta + \frac{p+1}{2})} K_{2, U_2, \zeta}^{-\alpha} f, \end{aligned}$$

where, $\Re(\alpha) > \frac{p-1}{2}, \Re(\zeta) > \frac{p-1}{2}$, $K_{2,U_2,\zeta}^{-\alpha}$ will denote the Erdélyi-Kober fractional integral operator of the second kind of order α and parameter ζ for the real matrix-variate case. Hence we have the following theorem:

Theorem 3.1 *When X_1 and X_2 are independently distributed $p \times p$ positive definite real matrix-variate random variables and when $X_2 = V$ and $U_2 = X_2^{\frac{1}{2}} X_1 X_2^{\frac{1}{2}}$ or $X_1 = V^{-\frac{1}{2}} U_2 V^{-\frac{1}{2}}, X_2 = V$ and when X_1 has a real matrix-variate type-1 beta distribution with the parameters $(\zeta + \frac{p+1}{2}, \alpha)$ and if $g_2(U_2)$ denotes the density of U_2 then*

$$\frac{\Gamma_p(\zeta+\frac{p+1}{2})}{\Gamma_p(\alpha+\zeta+\frac{p+1}{2})}g_2(U_2)=K_{2,U_2,\zeta}^{-\alpha}f \tag{3.10}$$

is Erdélyi-Kober fractional integral operator of the second kind of order α and parameter ζ for the real matrix-variate case, operating on f.

As a special case of (3.10), or independently, we can derive a result for the right-sided Weyl operator for the real matrix-variate case. Let the right-sided Weyl fractional integral operator of order α, parameter ζ and of the second kind or right-sided be denoted by $W_{2,X}^{-\alpha}$.

Theorem 3.2 *Let X_1, X_2, U_2, V be as defined in Theorem 3.1. Let X_1 have a type-1 beta density with the parameters $(\frac{p+1}{2}, \alpha)$. Let the density of X_2 be denoted by $f_4(X_2)=|X_2|^{\alpha}f(X_2)$ where $f(X_2)$ is arbitrary. Let the density of U_2 be denoted by $g_{21}(U_2)$. Then*

$$W_{2,U_2}^{-\alpha}f=\frac{1}{\Gamma_p(\alpha)}\int_{V>U_2>O}|V-U_2|^{\alpha-\frac{p+1}{2}}f(V)\mathrm{d}V=\frac{\Gamma_p(\frac{p+1}{2})}{\Gamma_p(\alpha+\frac{p+1}{2})}g_{21}(U_2),\ \Re(\alpha)>\frac{p-1}{2}. \tag{3.11}$$

The following notations will be used hereafter. We will denote a symmetric matrix product as $X_2^{\frac{1}{2}}X_1X_2^{\frac{1}{2}}=U_2$ and U_1 as a symmetric matrix ratio where $U_1=X_2^{\frac{1}{2}}X_1^{-1}X_2^{\frac{1}{2}}$. The density of U_2 will be denoted by $g_2(U_2)$ in general. Then for the j-th set of X_1 and X_2 the density of U_2 will be denoted as $g_{2j}(U_2)$. Similarly for the j-th set of X_1 and X_2 the density of U_1 will be denoted as $g_{1j}(U_1)$. For Erdélyi-Kober operators, letter K will be used, for Weyl operators letter W will be used and for Saigo and Caputo operators the letters S and C will be used respectively. For the first kind integral we will use the number 1 and for the second kind the number 2. D^{α} denotes the fractional differential operator of order α and $D^{-\alpha}$ the corresponding integral operator.

3.3 A Pathway Generalization of Erdélyi-Kober Fractional Integral Operator of the Second Kind in the Real Matrix-Variate Case

A pathway generalization, parallel to the results in the scalar case can be considered. For the general pathway models in real matrix-variate case, see Mathai [4]. In the pathway case when generalization to matrices is considered we take $\delta=1$. For a general δ, there will be problems with Jacobians of transformations for X^{δ} even if $X>O$ and $\delta>0$, see for example Mathai [3] for the case $\delta=2$ and when $X=X'$ to see the type of complications. Hence we take the case $\delta=1$ only. Let X_1 have a pathway density

$$f_5(X_1) = C_5|X_1|^{\gamma}|I - a(1-q)X|^{\frac{\eta}{1-q}} \tag{3.12}$$

for $I - a(1-q)X > O, q < 1, \eta > 0, a > 0$ where C_5 can be seen to be the following:

$$C_5 = \frac{[a(1-q)]^{p\gamma + \frac{p(p+1)}{2}} \Gamma_p(\gamma + \frac{\eta}{1-q} + (p+1))}{\Gamma_p(\gamma + \frac{p+1}{2})\Gamma_p(\frac{\eta}{1-q} + \frac{p+1}{2})}. \tag{3.13}$$

When $q > 1$ we may write $1 - q = -(q-1), q > 1$. Then the pathway density f_5 for X_1 changes to the form

$$f_7(X_1) = C_7|X_1|^{\gamma}|I + a(q-1)X|^{-\frac{\eta}{q-1}}, X_1 > O \tag{3.14}$$

for $a > 0, q > 1, \eta > 0, \Re(\gamma) > -1$ and

$$C_7 = \frac{[a(q-1)]^{p\gamma + \frac{p(p+1)}{2}} \Gamma_p(\frac{\eta}{q-1})}{\Gamma_p(\gamma + \frac{p+1}{2})\Gamma_p(\frac{\eta}{q-1} - \gamma - \frac{p+1}{2})} \tag{3.15}$$

for $\Re(\gamma) > -1, \Re(\frac{\eta}{q-1} - \gamma - \frac{p+1}{2}) > \frac{p-1}{2}$. The corresponding density for X_2 will be denoted by $f_8(X_2)$. When $q \to 1_-$ in (3.12) and $q \to 1_+$ in (3.14) the densities for X_1 will go to

$$f_9(X_1) = C_9|X_1|^{\gamma}e^{-\mathrm{tr}(a\eta X_1)}, X_1 > O \tag{3.16}$$

where $a > 0, \eta > 0$ and

$$C_9 = \frac{(a\eta)^{p\gamma + \frac{p(p+1)}{2}}}{\Gamma_p(\gamma + \frac{p+1}{2})} \tag{3.17}$$

for $a > 0, \eta > 0, \Re(\gamma) > -1$. The corresponding density for X_2 will be denoted by $f_{10}(X_2)$.

In (3.12) let X_2 have the density $f_8(X_2) = f(X_2)$ where f is arbitrary. Let X_1 and X_2 be statistically independently distributed. Let $U_2 = X_2^{\frac{1}{2}}X_1X_2^{\frac{1}{2}}, X_2 = V$ or $X_1 = V^{-\frac{1}{2}}U_2V^{-\frac{1}{2}}$. Let the density of U_2 be denoted as $g_{22}(U_2)$, corresponding to f_5 and f_6. Then, going through the earlier steps we have the following:

$$g_{22}(U_2) = c_5|U_2|^{\gamma}\int_{V > a(1-q)U_2 > O}|V|^{-\gamma-(\frac{\eta}{1-q}+\frac{p+1}{2})}|V - a(1-q)U_2|^{\frac{\eta}{1-q}}f(V)\mathrm{d}V. \tag{3.18}$$

Then

$$\Gamma_p(\gamma+\frac{p+1}{2})g_{22}(U_2)=\frac{[a(1-q)]^{p\gamma+\frac{p(p+1)}{2}}\Gamma_p(\gamma+\frac{\eta}{1-q}+\frac{(p+1)}{2})}{\Gamma_p(\frac{\eta}{1-q}+\frac{p+1}{2})}|U_2|^{\gamma}$$
$$\times\int_{V>a(1-q)U_2>O}|V-a(1-q)U_2|^{\frac{\eta}{1-q}}|V|^{-\gamma-(\frac{\eta}{1-q}+\frac{p+1}{2})}f(V)\mathrm{d}V$$
$$=K_{2,U_2,\gamma,a,q}^{-(\frac{\eta}{1-q}+\frac{p+1}{2})}f \tag{3.19}$$

where $K_{2,U_2,\gamma,a,q}^{-(\frac{\eta}{1-q}+\frac{p+1}{2})}f$ can be called the generalized pathway Erdélyi-Kober fractional integral operator of the second kind in the real matrix-variate case, operating on f. When the pathway parameter q varies from $-\infty$ to 1 it provides a pathway or a class of fractional integrals and all these fractional integrals in this pathway class will eventually go to the exponential form. For $a=1, q=0$, $\frac{\eta}{1-q}=\alpha-\frac{p+1}{2}$ and $\gamma=\zeta$ we have

$$K_{2,U_2,\gamma,a,q}^{-(\frac{\eta}{1-q}+\frac{p+1}{2})}f=K_{2,U_2,\zeta}^{-\alpha}f \tag{3.20}$$

the Erdélyi-Kober fractional integral of the second kind as a constant multiple of the density of the symmetric product of two matrix-variate independently distributed random variables. Note that when $q\to 1_-$ we can evaluate the limit of $g_2(U_2)$ by using the following lemmas:

Lemma 3.6

$$\lim_{q\to 1_-}c_5=\lim_{q\to 1_+}c_7=\frac{(a\eta)^{p\gamma+\frac{p(p+1)}{2}}}{\Gamma_p(\gamma+\frac{p+1}{2})}=c_9 \tag{i}$$

Proof Open up each $\Gamma_p(\cdot)$ in c_5 of (3.13) in terms of ordinary gamma functions. Then use the following asymptotic approximation for gamma functions. For $|z|\to\infty$ and γ a bounded quantity

$$\Gamma(z+\gamma)\approx\sqrt{2\pi}\,z^{z+\gamma-\frac{1}{2}}\mathrm{e}^{-z}. \tag{ii}$$

This is the first term in an asymptotic series for gamma functions. This term is also known as Stirling's approximation. When $q\to 1_-$ we have $\frac{1}{1-q}\to\infty$ and hence take $|z|$ as $\frac{\eta}{1-q}$ and expand all gammas by using Stirling's approximation to see that c_5 reduces to (i) above.

Lemma 3.7

$$\lim_{q\to 1_-}|I-a(1-q)X|^{\frac{\eta}{1-q}}=\mathrm{e}^{-a\eta\,\mathrm{tr}(X)}. \tag{iii}$$

Proof Writing the determinant in terms of eigenvalues we have

$$|I - a(1-q)X|^{\frac{\eta}{1-q}} = \prod_{j=1}^{p}(1 - a(1-q)\lambda_j)^{\frac{\eta}{1-q}} \tag{iv}$$

where $\lambda_1, \ldots, \lambda_p$ are the eigenvalues of X. Now

$$\lim_{q\to 1_-}(1 - a(1-q)\lambda_j)^{\frac{\eta}{1-q}} = \mathrm{e}^{-a\eta\,\lambda_j}. \tag{v}$$

Hence

$$\lim_{q\to 1_-}|I - a(1-q)X|^{\frac{\eta}{1-q}} = \prod_{j=1}^{p}\mathrm{e}^{-a\eta\,\lambda_j} = \mathrm{e}^{-a\eta\,\mathrm{tr}(X)}$$

which establishes (iii).

Now by using Lemmas 3.6 and 3.7 we have

$$\lim_{q\to 1_-} g_{22}(U_2) = \frac{(a\eta)^{p\gamma+\frac{p(p+1)}{2}}}{\Gamma_p(\gamma+\frac{p+1}{2})}|U_2|^{\gamma}\int_{V>O}|V|^{-\gamma-\frac{p+1}{2}}\mathrm{e}^{-a\eta\,\mathrm{tr}(V^{-\frac{1}{2}}U_1V^{-\frac{1}{2}})}\mathrm{d}V. \tag{3.21}$$

This is the limiting form of the pathway Erdélyi-Kober fractional integral of the second kind in this class of pathway fractional integrals of the second kind in the real matrix-variate case. Note that in the limiting situation, the fractional nature of the integral is lost.

In the pathway generalization, one can also replace the parameter a with a constant positive definite matrix A. In this case the model will be written as

$$f_{11}(X_1) = C_{11}(A)|X_1|^{\gamma}|I - (1-q)A^{\frac{1}{2}}X_1A^{\frac{1}{2}}|^{\frac{\eta}{1-q}} \tag{3.22}$$

for $q < 1, A > O, X_1 > O, I - (1-q)A^{\frac{1}{2}}X_1A^{\frac{1}{2}} > O$. The pathway parameter is still q. In this case

$$C_{11}(A) = \frac{(1-q)^{p\gamma+\frac{p(p+1)}{2}}|A|^{\gamma+\frac{p+1}{2}}\Gamma_p(\gamma+\frac{\eta}{1-q}+(p+1))}{\Gamma_p(\gamma+\frac{p+1}{2})\Gamma_p(\frac{\eta}{1-q}+\frac{p+1}{2})}. \tag{3.23}$$

Then $g_{22}(U_2)$ of (3.18) goes to the following form, denoted by $g_A(U_2)$

$$g_A(U_2) = C_{11}(A)|A|^{\frac{\eta}{1-q}}|U_2|^{\gamma}$$

$$\times \int_{V^*>O} |V^{\frac{1}{2}}A^{-1}V^{\frac{1}{2}} - (1-q)U_2|^{\frac{\eta}{1-q}} |V|^{-\gamma-(\frac{\eta}{1-q}+\frac{p+1}{2})} f(V)\mathrm{d}V \tag{3.24}$$

where

$$V^* = V^{\frac{1}{2}}A^{-1}V^{\frac{1}{2}} - (1-q)U_2.$$

Then one can define a pathway generalized Erdélyi-Kober fractional integral of the second kind in the real matrix-variate case as the following, taking the corresponding density as $f_{12}(X_2) = f(X_2)$ where f is arbitrary:

$$K_{2,U_2,\gamma,A,q}^{-(\frac{\eta}{1-q}+\frac{p+1}{2})} f = \Gamma_p(\gamma + \frac{p+1}{2}) g_A(U_2)$$

$$= \frac{(1-q)^{p\gamma+\frac{p(p+1)}{2}} |A|^{\gamma+\frac{\eta}{1-q}+\frac{p+1}{2}} \Gamma_p(\gamma + \frac{\eta}{1-q} + (p+1))}{\Gamma_p(\frac{\eta}{1-q} + \frac{p+1}{2})} |U_2|^{\gamma}$$

$$\times \int_{V^*>O} |V^{\frac{1}{2}}A^{-1}V^{\frac{1}{2}} - (1-q)U_2|^{\frac{\eta}{1-q}} |V|^{-\gamma-(\frac{\eta}{1-q}+\frac{p+1}{2})} f(V)\mathrm{d}V \tag{3.25}$$

In this case, as $q \to 1_-$ we have

$$\lim_{q\to 1_-} g_A(U_2) = \frac{|A|^{\gamma+\frac{p+1}{2}} \eta^{p\gamma+\frac{p(p+1)}{2}}}{\Gamma_p(\gamma + \frac{p+1}{2})} |U_2|^{\gamma}$$

$$\times \int_{V>A} |V|^{-\gamma-\frac{p+1}{2}} \mathrm{e}^{-\eta\, \mathrm{tr}(A^{\frac{1}{2}}V^{-\frac{1}{2}}U_2V^{-\frac{1}{2}}A^{\frac{1}{2}})} f(V)\mathrm{d}V. \tag{3.26}$$

3.4 M-Transforms of Erdélyi-Kober Fractional Integral of the Second Kind in the Real Matrix-Variate Case

The generalized matrix transform or M-transform is defined and illustrated in Mathai [3]. The M-transform of Erdélyi-Kober fractional integral of the second kind in the real matrix-variate case is the following:

Theorem 3.3 *For the Erdélyi-Kober fractional integral of the second kind in the real matrix-variate case defined in* (3.9) *the M-transform with parameter s is given by*

$$M\{K_{2,X,\zeta}^{-\alpha}f;s\}=\int_{X>O}|X|^{s-\frac{p+1}{2}}[\int_{T>X>O}\frac{|X|^{\zeta}}{\Gamma_p(\alpha)}|T-X|^{\alpha-\frac{p+1}{2}}|T|^{-\zeta-\alpha}f(T)\mathrm{d}T]\mathrm{d}X$$
$$=\frac{\Gamma_p(\zeta+s)}{\Gamma(\alpha+\zeta+s)}f^*(s),\ \Re(\zeta+s)>\frac{p-1}{2},\Re(\alpha)>\frac{p-1}{2} \quad (3.27)$$

where $f^*(s)$ is the M-transform of $f(X)$.

Proof Interchanging the integral we have

$$M\{K_{2,X,\zeta}^{-\alpha}f;s\}=\int_{T>O}|T|^{-\zeta-\alpha}f(T)[\frac{1}{\Gamma_p(\alpha)}\int_{O<X<T}|X|^{\zeta+s-\frac{p+1}{2}}|T-X|^{\alpha-\frac{p+1}{2}}\mathrm{d}X]\mathrm{d}T.$$

Note that

$$|T-X|=|T|\,|I-T^{-\frac{1}{2}}XT^{-\frac{1}{2}}|,\ Y=T^{-\frac{1}{2}}XT^{-\frac{1}{2}}\Rightarrow \mathrm{d}Y=|T|^{-\frac{p+1}{2}}\mathrm{d}X.$$

Hence

$$\int_{O<X<T}|X|^{\zeta-\frac{p+1}{2}}|T-X|^{\alpha-\frac{p+1}{2}}\mathrm{d}X=|T|^{\alpha+\zeta+s-\frac{p+1}{2}}\int_Y|Y|^{\zeta+s-\frac{p+1}{2}}|I-Y|^{\alpha-\frac{p+1}{2}}\mathrm{d}Y.$$

We can evaluate the Y-integral by using real matrix-variate type-1 beta integral.

$$\int_{O<Y<I}|Y|^{\zeta+s-\frac{p+1}{2}}|I-Y|^{\alpha-\frac{p+1}{2}}\mathrm{d}Y=\frac{\Gamma_p(\zeta+s)\Gamma_p(\alpha)}{\Gamma_p(\alpha+\zeta+s)}$$

for $\Re(\alpha)>\frac{p-1}{2}$, $\Re(\zeta+s)>\frac{p-1}{2}$. Now the T-integral gives

$$\int_{T>O}|T|^{s-\frac{p+1}{2}}f(T)\mathrm{d}T=f^*(s)$$

where $f^*(s)$ is the M-transform of $f(X)$. Hence (3.27) follows. Note that for $p=1$ the result agrees with that in the scalar case, which is available in the literature, see for example Mathai and Haubold [10].

From (3.27) for $\zeta=0$ and $\Re(\alpha)>\frac{p-1}{2}$ we have the special case of the Erdélyi-Kober fractional integral of the second kind in the real matrix-variate case

$$K_{2,X,0}^{-\alpha}f=\frac{1}{\Gamma(\alpha)}\int_{T>X>O}|T-X|^{\alpha-\frac{p+1}{2}}|T|^{-\alpha}f(T)\mathrm{d}T. \quad (3.28)$$

But the right side of (3.28) is Weyl fractional integral of the second kind of order α in the matrix-variate case, ${}_XW_{\infty}^{-\alpha}f=W_{2,X}^{-\alpha}f$, except for the factor $|T|^{-\alpha}$. The Weyl integral of the second kind in the real matrix case is

$$ {}_XW_{\infty}^{-\alpha}f(X) = W_{2,X}^{-\alpha}f = \frac{1}{\Gamma_p(\alpha)}\int_{T>X>O}|T-X|^{\alpha-\frac{p+1}{2}}f(T)\mathrm{d}T, \Re(\alpha) > \frac{p-1}{2}. \tag{3.29}$$

The ∞ sitting in ${}_XW_{\infty}^{-\alpha}$ is to be interpreted as that X is not bounded above by a positive definite constant matrix. Hence we have the following corollary.

Corollary 3.1 *The M-transform of the right-sided or second kind Weyl fractional integral in the real matrix-variate case is given by*

$$ M\{{}_XW_{\infty}^{-\alpha}|X|^{-\alpha}f(X);s\} = M\{W_{2,X}^{-\alpha}|X|^{-\alpha}f\} = \frac{\Gamma_p(s)}{\Gamma_p(\alpha+s)}f^*(s) \tag{3.30}$$

for $\Re(s) > \frac{p-1}{2}, \Re(\alpha) > \frac{p-1}{2}$ *where* $f^*(s)$ *is the M-transform of* $f(X)$.

The proof is parallel to that in Theorem 3.3. Let us see whether a Mellin convolution type formula holds for Erdélyi-Kober fractional integral of the second kind in the matrix-variate case. Let

$$ g_{23}(U_2) = \int_V |V|^{-\frac{p+1}{2}}f_{13}(V^{-\frac{1}{2}}U_2V^{-\frac{1}{2}})f_{14}(V)\mathrm{d}V \tag{3.31}$$

where $f_{13}(X_1)$ is a type-1 matrix-variate beta density with parameters $(\zeta+\frac{p+1}{2},\alpha)$. That is,

$$ f_{13}(X_1) = \frac{\Gamma_p(\alpha+\zeta+\frac{p+1}{2})}{\Gamma_p(\alpha)\Gamma_p(\zeta+\frac{p+1}{2})}|X_1|^{\zeta}|I-X_1|^{\alpha-\frac{p+1}{2}}, O<X_1<I \tag{3.32}$$

for $\Re(\alpha) > \frac{p-1}{2}, \Re(\zeta) > -1$ and $f_{13}(X_1) = 0$ elsewhere. Let $f_{14}(X_2) = f(X_2)$ be the corresponding density for X_2 where f is arbitrary. Substituting (3.32) in (3.31) we have

$$\begin{aligned}\frac{\Gamma_p(\zeta+\frac{p+1}{2})}{\Gamma_p(\alpha+\zeta+\frac{p+1}{2})}g_{23}(U_2) &= \frac{1}{\Gamma_p(\alpha)}\int_V|V|^{-\frac{p+1}{2}}|U_2|^{\zeta}|V|^{-\zeta}|I-V^{-\frac{1}{2}}U_2V^{-\frac{1}{2}}|^{\alpha-\frac{p+1}{2}}f(V)\mathrm{d}V\\ &= \frac{|U_2|^{\zeta}}{\Gamma_p(\alpha)}\int_V|V|^{-\zeta-\alpha}|V-U_2|^{\alpha-\frac{p+1}{2}}f(V)\mathrm{d}V\\ &= \frac{|U_2|^{\zeta}}{\Gamma_p(\alpha)}\int_{V>U_2>O}|V-U_2|^{\alpha-\frac{p+1}{2}}|V|^{-\zeta-\alpha}f(V)\mathrm{d}V\\ &= K_{2,U_2,\zeta}^{-\alpha}f \end{aligned}\tag{3.33}$$

which is the Erdélyi-Kober fractional integral of the second kind of order α in the real matrix-variate case. Hence we have the following theorem:

Theorem 3.4 *A constant multiple of the Erdélyi-Kober fractional integral of the second kind of order α in the real matrix-variate case can also be represented as a Mellin convolution type formula*

$$K_{2,X,\zeta}^{-\alpha}f = \int_V |V|^{-\frac{p+1}{2}} f_{13}(V^{-\frac{1}{2}}XV^{-\frac{1}{2}})f_{14}(V)\mathrm{d}V$$

where $f_{13}(X_1)$ is a type-1 beta density with parameters $(\zeta + \frac{p+1}{2}, \alpha)$ and $f_2(V) = f(V)$ is an arbitrary function or arbitrary density if the Erdélyi-Kober fractional integral is to be taken as a constant multiple of a statistical density.

3.5 Generalization in Terms of Hypergeometric Series for Erdélyi-Kober Fractional Integral of the Second Kind in the Real Matrix-Variate Case

For introducing hypergeometric series of matrix argument we will need the definitions, notation and lemmas. Hypergeometric functions of matrix argument are defined in terms of matrix-variate Laplace transforms, M-transforms and zonal polynomials. Explicit series form for all cases is available through the definition in terms of zonal polynomials and hence we will define in terms of zonal polynomials. Some derivations and properties of zonal polynomials may be seen from Mathai, Provost and Hayakawa [15].

$$_rF_s(Z) = {_rF_s}(a_1, \ldots, a_r; b_1, \ldots, b_s; Z) = \sum_{k=0}^{\infty}\sum_K \frac{(a)_K \ldots (a_r)_K}{(b_1)_K \ldots (b_s)_K}\frac{C_K(Z)}{k!} \tag{3.34}$$

where $K = (k_1, \ldots, k_p), k_1 + \ldots + k_p = k$ is a partition of $k = 0, 1, 2, \ldots$

$$(a)_K = \prod_{j=1}^{p}(a - \frac{j-1}{2})_{k_j}, (b)_{k_j} = b(b+1)\ldots(b+k_j-1), (b)_0 = 1, b \neq 0 \tag{3.35}$$

and $C_K(Z)$ is a zonal polynomial of order k and Z is a $p \times p$ matrix. The series is defined for the real and complex matrices. Zonal polynomials are certain symmetric functions of the eigenvalues of Z. In our discussions, Z will be real and positive definite. For more details about zonal polynomials see Mathai et al. [15]. The following basic results are needed in our discussions. A standard notation in this area is

$$\Gamma_p(\alpha, K) = \Gamma_p(\alpha)(\alpha)_K. \tag{3.36}$$

The following basic results are needed in our discussion.

Lemma 3.8

$$\int_O^I |X|^{\alpha-\frac{p+1}{2}}|I-X|^{\beta-\frac{p+1}{2}}C_K(TX)\mathrm{d}X = \frac{\Gamma_p(\alpha,K)\Gamma_p(\beta)}{\Gamma_p(\alpha+\beta,K)}C_K(T) \tag{3.37}$$

for $\Re(\alpha) > \frac{p-1}{2}, \Re(\beta) > \frac{p-1}{2}$.

Lemma 3.9 *For* $\Re(\alpha) > \frac{p-1}{2}$, $A > O, S > O$

$$\int_{O<S<A} |S|^{\alpha-\frac{p+1}{2}}C_K(ZS)\mathrm{d}S = \frac{\Gamma_p(\alpha,K)\Gamma_p(\frac{p+1}{2})}{\Gamma_p(\alpha+\frac{p+1}{2},K)}|A|^{\alpha}C_K(ZA). \tag{3.38}$$

Let us assume that all the parameters $a_1,\ldots,a_r,b_1,\ldots,b_s$ are real and positive and let the argument matrices be $p\times p$ and positive definite. For $A > O$, let the density of X_1 be

$$\begin{aligned} f_{15}(X_1) &= \frac{1}{c_f}{}_rF_s(a_1,\ldots,a_r;b_1,\ldots,b_s;AX_1)|X_1|^{\zeta}|I-X_1|^{\alpha-\frac{p+1}{2}} \\ &= \frac{1}{c_f}\sum_{k=0}^{\infty}\sum_K \frac{(a_1)_K\ldots(a_r)_K}{(b_1)_K\ldots(b_s)_K}\frac{1}{k!}C_K(AX_1)|X_1|^{\zeta}|I-X_1|^{\alpha-\frac{p+1}{2}} \end{aligned} \tag{3.39}$$

where the normalizing constant $\frac{1}{c_f}$ is available by integrating out term by term with the help of Lemma 3.8. It will be available in terms of a ${}_{r+1}F_{s+1}$. Let the corresponding density for X_2, $f_{16}(X_2) = f(X_2)$ be an arbitrary density. As before, let $U_2 = X_2^{\frac{1}{2}}X_1X_2^{\frac{1}{2}}$, $X_2 = V$, or $X_1 = V^{-\frac{1}{2}}U_2V^{-\frac{1}{2}}$, then denoting the density of U_2, denoted by $g_{24}(U_2)$, we have

$$\begin{aligned} g_{24}(U_2) &= \int_V f_{15}(V^{-\frac{1}{2}}U_2V^{-\frac{1}{2}})f(V)|V|^{-\frac{p+1}{2}}\mathrm{d}V \\ &= \frac{1}{c_f}\frac{\Gamma_p(\alpha+\zeta+\frac{p+1}{2})}{\Gamma_p(\zeta+\frac{p+1}{2})\Gamma_p(\alpha)}\sum_{k=0}^{\infty}\sum_K\frac{(a_1)_K\ldots(a_r)_K}{(b_1)_K\ldots(b_s)_K}\frac{1}{k!} \\ &\times\int_V |V^{-\frac{1}{2}}U_2V^{-\frac{1}{2}}|^{\zeta}|I-V^{-\frac{1}{2}}U_2V^{-\frac{1}{2}}|^{\alpha-\frac{p+1}{2}}|V|^{-\frac{p+1}{2}} \\ &\times C_K(AV^{-\frac{1}{2}}U_2V^{-\frac{1}{2}})f(V)\mathrm{d}V \end{aligned} \tag{3.40}$$

This is the generalization of a constant times the Erdélyi-Kober fractional integral of order α and parameter ζ of the second kind in the real matrix-variate case. For ${}_rF_s = {}_2F_1$ one has the matrix-variate generalization of a constant times the Saigo operator of the second kind in the real matrix-variate case.

3.6 Erdélyi-Kober Fractional Integral of the First Kind in the Real Matrix-Variate Case

Definition 3.2 We will give the following definition and notation for Erdélyi-Kober fractional integral of the first kind in the real matrix-variate case:

$$K_{1,X,\zeta}^{-\alpha} f = \frac{|X|^{-\zeta-\alpha}}{\Gamma_p(\alpha)} \int_{O<V<X} |X - V|^{\alpha - \frac{p+1}{2}} |V|^{\zeta} f(V) \mathrm{d}V \tag{3.41}$$

for $\Re(\zeta) > -1, \Re(\alpha) > \frac{p-1}{2}$. This definition is used because, for $p = 1$ in the real scalar variable case the corresponding item is called Erdélyi-Kober fractional integral of the first kind.

Theorem 3.5 *For $\Re(\alpha) > \frac{p-1}{2}, \Re(\zeta) > -1$ the M-transform, with parameter s, of Erdélyi-Kober fractional integral of the first kind of order α and parameter ζ in the real matrix-variate case, is given by*

$$\begin{aligned} M\{K_{1,X,\zeta}^{-\alpha} f; s\} &= \int_{X>O} |X|^{s-\frac{p+1}{2}} [\frac{|X|^{-\zeta-\alpha}}{\Gamma_p(\alpha)} \int_{O<V<X} |X-V|^{\alpha-\frac{p+1}{2}} |V|^{\zeta} f(V) \mathrm{d}V] \mathrm{d}X \\ &= \frac{\Gamma_p(\zeta + \frac{p+1}{2} - s)}{\Gamma_p(\alpha + \zeta + \frac{p+1}{2} - s)} f^*(s), \\ &\Re(s) < \Re(\zeta + 1), \Re(\alpha) > \frac{p-1}{2} \end{aligned} \tag{3.42}$$

where $f^*(s)$ is the M-transform of $f(X)$.

Proof Integrating out X first we have the X-integral

$$\begin{aligned} &\int_{X>V>O} |X|^{s-\zeta-\alpha-\frac{p+1}{2}} |X - V|^{\alpha-\frac{p+1}{2}} \mathrm{d}X \\ &= \int_{Y>O} |Y + V|^{s-\zeta-\alpha-\frac{p+1}{2}} |Y|^{\alpha-\frac{p+1}{2}} \mathrm{d}Y, \ (Y = X - V) \\ &= |V|^{s-\zeta-\alpha-\frac{p+1}{2}} \int_{Y>O} |I + V^{-\frac{1}{2}} Y V^{-\frac{1}{2}}|^{s-\zeta-\alpha-\frac{p+1}{2}} |Y|^{\alpha-\frac{p+1}{2}} \mathrm{d}Y. \end{aligned}$$

Put $Z = V^{-\frac{1}{2}} Y V^{-\frac{1}{2}} \Rightarrow \mathrm{d}Z = |V|^{-\frac{p+1}{2}} \mathrm{d}Y$. Then the X-integral is

$$|V|^{s-\zeta-\frac{p+1}{2}} \int_{Z>O} |Z|^{\alpha-\frac{p+1}{2}} |I + Z|^{-(\frac{p+1}{2}+\alpha+\zeta-s)} \mathrm{d}Z = |V|^{s-\zeta-\frac{p+1}{2}} \frac{\Gamma_p(\alpha)\Gamma_p(\frac{p+1}{2} + \zeta - s)}{\Gamma_p(\frac{p+1}{2} + \alpha + \zeta - s)}$$

for $\Re(\alpha) > \frac{p-1}{2}, \Re(\zeta - s) > -1$ by evaluating the integral by using a type-2 matrix-variate beta integral in the real case. Now, the V-integral becomes

$$\int_{V>O} |V|^{s-\frac{p+1}{2}} f(V)\mathrm{d}V = f^*(s).$$

Hence

$$M\{K_{1,X,\zeta}^{-\alpha} f; s\} = \frac{\Gamma_p(\frac{p+1}{2} + \zeta - s)}{\Gamma_p(\frac{p+1}{2} + \alpha + \zeta - s)} f^*(s) \tag{3.43}$$

for $\Re(\alpha) > \frac{p-1}{2}, \Re(\zeta - s) > -1$. Note that for $\zeta = 0$,

$$K_{1,X,0}^{-\alpha} f = |X|^{-\alpha} {}_0D_X^{-\alpha} f = |X|^{-\alpha} D_{1,X}^{-\alpha} f \tag{3.44}$$

where ${}_0D_X^{-\alpha}$ is the left-sided Riemann-Liouville fractional integral of order α for the real matrix-variate case. Note that for the scalar case, for $p = 1$,

$$M\{K_{1,x,\zeta}^{-\alpha} f; s\} = \frac{\Gamma(1 + \zeta - s)}{\Gamma(1 + \alpha + \zeta - s)} \tag{3.45}$$

for $\Re(\alpha) > 0, \Re(\zeta - s) > -1$ agreeing with the corresponding Mellin transform in the scalar case.

Corollary 3.2 *The M-transform of* $|X|^{-\alpha} {}_0D_X^{-\alpha} f = |X|^{-\alpha} D_{1,X}^{-\alpha} f$ *is given by*

$$M\{|X|^{-\alpha} {}_0D_X^{-\alpha} f; s\} = M\{|X|^{-\alpha} D_{1,X}^{-\alpha} f; s\} = \frac{\Gamma_p(\frac{p+1}{2} - s)}{\Gamma_p(\frac{p+1}{2} + \alpha - s)} f^*(s) \tag{3.46}$$

for $\Re(\alpha) > \frac{p-1}{2}, \Re(s) < 1$.

The proof is parallel to that in Theorem 3.5.

Let us treat a Erdélyi-Kober fractional integral operator of the first kind operating on f as a statistical density. Let X_2 have a real matrix-variate density $f_{18}(X_2) = f(X_2)$, where f is arbitrary, and let X_1 have a real matrix-variate type-1 beta density $f_{17}(X_1)$ with parameters (ζ, α). That is,

$$f_{17}(X_1) = \frac{\Gamma_p(\zeta + \alpha)}{\Gamma_p(\zeta)\Gamma_p(\alpha)} |X_1|^{\zeta - \frac{p+1}{2}} |I - X_1|^{\alpha - \frac{p+1}{2}}, O < X_1 < I \tag{3.47}$$

for $\Re(\zeta) > \frac{p-1}{2}, \Re(\alpha) > \frac{p-1}{2}$ and $f_{17}(X_1) = 0$ elsewhere. Let X_1 and X_2 be statistically independently distributed. Let $U_1 = X_2^{\frac{1}{2}} X_1^{-1} X_2^{\frac{1}{2}}$ be the symmetric ratio of X_2 to X_1. Consider the transformation $X_2 = V, X_1 = V^{\frac{1}{2}} U_1^{-1} V^{\frac{1}{2}}$. The Jacobian is given by

$$\mathrm{d}X_1 \wedge \mathrm{d}X_2 = |V|^{\frac{p+1}{2}} |U_1|^{-(p+1)} \mathrm{d}U_1 \wedge \mathrm{d}V.$$

The marginal density of U_1, denoted by $g_{11}(U_1)$, is given by

$$\begin{aligned} g_{11}(U_1) &= \frac{\Gamma_p(\zeta+\alpha)}{\Gamma_p(\zeta)\Gamma_p(\alpha)} \int_V |V^{\frac{1}{2}} U_1^{-1} V^{\frac{1}{2}}|^{\zeta - \frac{p+1}{2}} \\ &\quad \times |I - V^{\frac{1}{2}} U_1^{-1} V^{\frac{1}{2}}|^{\alpha - \frac{p+1}{2}} f(V) |V|^{\frac{p+1}{2}} |U_1|^{-(p+1)} \mathrm{d}V \\ &= \frac{\Gamma_p(\zeta+\alpha)}{\Gamma_p(\zeta)\Gamma_p(\alpha)} |U_1|^{-\zeta-\alpha} \int_{O<V<U_1} |U_1 - V|^{\alpha - \frac{p+1}{2}} |V|^{\zeta} f(V) \mathrm{d}V \\ &= \frac{\Gamma_p(\zeta+\alpha)}{\Gamma_p(\zeta)} K_{1,U_1,\zeta}^{-\alpha} f, \end{aligned}$$

where $K_{1,U_1,\zeta}^{-\alpha}$ will denote the Erd'elyi-Kober fractional integral operator of the first kind and of order α and parameter ζ. Therefore

$$\frac{\Gamma_p(\zeta)}{\Gamma_p(\zeta+\alpha)} g_{11}(U_1) = \frac{|U_1|^{-\zeta-\alpha}}{\Gamma_p(\alpha)} \int_{O<V<U} |V|^{\zeta} f(V) \mathrm{d}V = K_{1,U_1,\zeta}^{-\alpha} f. \tag{3.48}$$

This is Erdélyi-Kober fractional integral of the first kind of order α and parameter ζ in the real matrix-variate case and it can be considered as a constant multiple of a real matrix-variate statistical density. For $p = 1$ the integral in (3.48) is Erdélyi-Kober fractional integral of the first kind of order α and parameter ζ in the real scalar variable case and hence we will call the fractional integral in (3.48) as the corresponding fractional integral in the real matrix-variate case.

One can also consider a pathway extension for the real matrix-variate Erdélyi-Kober fractional integral operator of the first kind in the real matrix-variate case.

3.7 Pathway Extension of Erdélyi-Kober Fractional Integral of the First Kind in the Real Matrix-Variate Case

Consider the following pathway modified form of the density for X_1. That is,

$$f_{19}(X_1) = C_{19} |X_1|^{\gamma - \frac{p+1}{2}} |I - a(1-q)X_1|^{\frac{\eta}{1-q}}, \ I - a(1-q)X_1 > O \tag{3.49}$$

for $q < 1, a > 0, \eta > 0$ where

$$C_{19} = \frac{[a(1-q)]^{p\gamma} \Gamma_p(\gamma + \frac{\eta}{1-q} + \frac{p+1}{2})}{\Gamma_p(\frac{\eta}{1-q} + \frac{p+1}{2}) \Gamma_p(\gamma)}, \tag{3.50}$$

and the corresponding density X_2, $f_{20}(X_2) = f(X_2)$ where f is arbitrary. Consider the same type of transformation as before: $X_2 = V, X_1 = V^{\frac{1}{2}}U_1^{-1}V^{\frac{1}{2}}$. The marginal density of U_1, denoted by $g_{12}(U_1)$, is given by

$$\begin{aligned} g_{12}(U_1) &= C_{19}\int_V |V^{\frac{1}{2}}U_1^{-1}V^{\frac{1}{2}}|^{\gamma-\frac{p+1}{2}}|I-a(1-q)V^{\frac{1}{2}}U_1^{-1}V^{\frac{1}{2}}|^{\frac{\eta}{1-q}} \\ &\quad\times f(V)|V|^{\frac{p+1}{2}}|U_1|^{-(p+1)}\mathrm{d}V \qquad (3.51) \\ &= C_{19}|U_1|^{-\gamma-(\frac{\eta}{1-q}+\frac{p+1}{2})}\int_{U_1>a(1-q)V>O}|U_1-a(1-q)V|^{\frac{\eta}{1-q}}|V|^{\gamma}f(V)\mathrm{d}V. \end{aligned}$$

Then

$$\begin{aligned} \Gamma_p(\gamma)g_{12}(U_1) &= \frac{[a(1-q)]^{p\gamma}\Gamma_p(\gamma+\frac{\eta}{1-q}+\frac{p+1}{2})}{\Gamma_p(\frac{\eta}{1-q}+\frac{p+1}{2})}|U_1|^{-\gamma-(\frac{\eta}{1-q}+\frac{p+1}{2})} \\ &\quad\times\int_{U_1>a(1-q)V>O}|U_1-a(1-q)V|^{\frac{\eta}{1-q}}|V|^{\gamma}f(V)\mathrm{d}V. \qquad (3.52) \end{aligned}$$

The right side of (3.52) is the pathway extension of Erdélyi-Kober fractional integral of the first kind in the real matrix-variate case. The right side divided by $\Gamma_p(\gamma)$ is also a statistical density of a type of ratio of independently distributed matrix-variate random variables.

Note that for $a = 1, q = 0, \frac{\eta}{1-q} + \frac{p+1}{2} = \alpha$, (3.52) reduces to the special case (3.47) for $\gamma = \zeta$. Thus, (3.52) describes a vast family of fractional integrals which can all be considered as generalizations of the Erdélyi-Kober fractional integral of the first kind in the real matrix-variate case. The limiting form when $q \to 1_-$ is available from the structure in (3.51). Note that

$$\lim_{q\to 1_-}|I-a(1-q)V^{\frac{1}{2}}U_1^{-1}V^{\frac{1}{2}}|^{\frac{\eta}{1-q}} = \mathrm{e}^{-a\eta\,\mathrm{tr}(V^{\frac{1}{2}}U_1^{-1}V^{\frac{1}{2}})} \qquad (3.53)$$

Hence

$$\begin{aligned} \lim_{q\to 1_-} g_{12}(U_1) &= (\lim_{q\to 1_-} C_{19})\int_{V>O}|U_1|^{-\gamma-\frac{p+1}{2}}|V|^{\gamma} \\ &\quad\times\mathrm{e}^{-a\eta\,\mathrm{tr}(V^{\frac{1}{2}}U_1^{-1}V^{\frac{1}{2}})}f(V)\mathrm{d}V \qquad (3.54) \end{aligned}$$

where

$$\lim_{q\to 1_-} C_{19} = \frac{(a\eta)^{p\gamma}}{\Gamma_p(\gamma)}.$$

That is,

$$\lim_{q\to 1_-} g_{12}(U_1)=\frac{(a\eta)^{p\gamma}}{\Gamma_p(\gamma)}|U_1|^{-\gamma-\frac{p+1}{2}}\int_{V>O}|V|^{\gamma}\mathrm{e}^{-a\eta\,\mathrm{tr}(V^{\frac{1}{2}}U_1^{-1}V^{\frac{1}{2}})}f(V)\mathrm{d}V=g_{13}(U_1). \tag{3.55}$$

In this case also one can replace the parameter a in $f_{19}(X_1)$ by a constant positive definite matrix $A>O$. Then $f_{19}(X_1)$ denoted by $f_{21}(X_1)$ can be written as

$$f_{21}(X_1)=C_{21}(A)|X_1|^{\gamma-\frac{p+1}{2}}|I-(1-q)A^{\frac{1}{2}}X_1A^{\frac{1}{2}}|^{\frac{\eta}{1-q}}$$

where

$$C_{21}(A)=\frac{(1-q)^{p\gamma}|A|^{\gamma}\Gamma_p(\gamma+\frac{\eta}{1-q}+\frac{p+1}{2})}{\Gamma_p(\gamma)\Gamma_p(\frac{\eta}{1-q}+\frac{p+1}{2})}.$$

Then the density of U_1, where $X_1=V^{\frac{1}{2}}U_1^{-1}V^{\frac{1}{2}}$, $X_2=V$, denoted by $g_{14}(U_1)$, is given by

$$g_{14}(U_1)=C_{21}(A)|U_1|^{-\gamma-(\frac{\eta}{1-q}+\frac{p+1}{2})}\int_{U_1>(1-q)V^{\frac{1}{2}}AV^{\frac{1}{2}}>O}|V|^{\gamma}|U_1-(1-q)V^{\frac{1}{2}}AV^{\frac{1}{2}}|^{\frac{\eta}{1-q}}f(V)\mathrm{d}V. \tag{3.56}$$

3.8 A General Definition

From the various types of fractional integrals of order α, $\Re(\alpha)>\frac{p-1}{2}$, defined so far for the real $p\times p$ matrix-variate case a few observations can be made. They are all M-convolutions of products and ratios where one function $f_1(X_1)$ is of the form $f_1(X_1)=\phi_1(X_1)\frac{1}{\Gamma_p(\alpha)}|I-X_1|^{\alpha-\frac{p+1}{2}}$ and the other function is of the form $f_2(X_2)=\phi_2(X_2)f(X_2)$ where f is an arbitrary function and $\phi_1(X_1)$ and $\phi_2(X_2)$ are some specified functions. We can make use of this observation and define fractional integrals of the first kind and second kind as follows: Let

$$f_1(X_1)=\phi_1(X_1)\frac{1}{\Gamma_p(\alpha)}|I-X_1|^{\alpha-\frac{p+1}{2}} \text{ and } f_2(X_2)=\phi_2(X_2)f(X_2) \tag{3.57}$$

Let $U_2=X_2^{\frac{1}{2}}X_1X_2^{\frac{1}{2}}$ with $X_2=V$ or $X_1=V^{-\frac{1}{2}}U_2V^{-\frac{1}{2}}$ be the symmetric product and $U_1=X_2^{\frac{1}{2}}X_1^{-1}X_2^{\frac{1}{2}}$ with $X_2=V$ or $X_1=V^{\frac{1}{2}}U_1^{-1}V^{\frac{1}{2}}$ be the symmetric ratio of X_2 to X_1. Then the Jacobians are already evaluated. They are

$$\mathrm{d}X_1\wedge\mathrm{d}X_2=|V|^{-\frac{p+1}{2}}\mathrm{d}U_2\wedge\mathrm{d}V=|V|^{\frac{p+1}{2}}|U_1|^{-(p+1)}\mathrm{d}U_1\wedge\mathrm{d}V.$$

Let the density of U_2, if f_1 and f_2 are statistical densities and if not let the M-convolution of the product be denoted by $g_2(U_2)$, and the density of U_1 or the M-convolution of a ratio be denoted by $g_1(U_1)$.

Definition 3.3 (Fractional integral of the second kind of order α in the real matrix-variate case) Fractional integrals of the second kind of order α in the real positive definite $p \times p$ matrix-variate case is defined as $g_2(U_2)$ where,

$$g_2(U_2)=\frac{1}{\Gamma_p(\alpha)}\int_V |V|^{-\frac{p+1}{2}}\phi_1(V^{-\frac{1}{2}}U_2V^{-\frac{1}{2}})|I-V^{-\frac{1}{2}}U_2V^{-\frac{1}{2}}|^{\alpha-\frac{p+1}{2}}\phi_2(V)f(V)\mathrm{d}V. \tag{3.58}$$

3.8.1 Special Cases

Case (1): Let $\phi_1(X_1) = |X_1|^{\gamma} = |V^{-\frac{1}{2}}U_2V^{-\frac{1}{2}}|^{\gamma}, \phi_2 = 1, \Re(\alpha) > \frac{p-1}{2}$. Then (3.58) becomes

$$\frac{|U_2|^{\gamma}}{\Gamma_p(\alpha)}\int_{V>U_2>O}|V|^{-\gamma-\alpha}|V-U_2|^{\alpha-\frac{p+1}{2}}f(V)\mathrm{d}V = K_{2,U_2,\gamma}^{-\alpha}f \tag{3.59}$$

which is Erdélyi-Kober fractional integral of order α, parameter γ and of the second kind in the real $p \times p$ matrix-variate case.

Case (2): Let $\phi_1 = 1, \phi_2(V) = |V|^{\alpha}, \Re(\alpha) > \frac{p-1}{2}$. Then (3.58) becomes

$$\frac{1}{\Gamma_p(\alpha)}\int_{V>U_2>O}|V-U_2|^{\alpha-\frac{p+1}{2}}f(V)\mathrm{d}V = W_{2,U_2}^{-\alpha}f \tag{3.60}$$

which is Weyl fractional integral of order α and of the second kind in the real $p \times p$ matrix-variate case. If there is an upper bound B for V, where $B > O$ is a constant positive definite matrix, then (3.60) is Riemann-Liouville fractional integral of the second kind of order α in the real $p \times p$ matrix-variate case. By specializing ϕ_1 and ϕ_2 one can derive all the fractional integrals of the second kind in the literature for $p = 1$ and hence the corresponding $g_2(U_2)$ of (3.58) can be taken as the corresponding real matrix-variate cases.

Definition 3.4 (Fractional integral of the first kind of order α in the real matrix-variate case) Fractional integrals of the first kind of order α in the real positive definite $p \times p$ matrix-variate case is defined as $g_1(U_1)$, the density of the symmetric ratio of matrices X_2 to X_1 or M-convolution of a ratio as given above, where

$$g_1(U_1) = \frac{1}{\Gamma_p(\alpha)} \times \int_V |V|^{\frac{p+1}{2}}|U_1|^{-(p+1)}\phi_1(V^{\frac{1}{2}}U_1^{-1}V^{\frac{1}{2}})|I-V^{\frac{1}{2}}U_1^{-1}V^{\frac{1}{2}}|^{\alpha-\frac{p+1}{2}}\phi_2(V)f(V)\mathrm{d}V, \tag{3.61}$$

for $\Re(\alpha) > \frac{p-1}{2}$. One should be able to get all fractional integrals of order α and of the first kind from (3.61) by specializing ϕ_1 and ϕ_2.

3.8.2 Special Cases of First Kind Fractional Integrals

Case (1): Let $\phi_1(X_1) = |X_1|^{\gamma-\frac{p+1}{2}}, \phi_2 = 1$ in (3.61) for $\Re(\alpha) > \frac{p-1}{2}$. Then $g_1(U_1)$ of (3.61) becomes the following:

$$\frac{|U_1|^{-\alpha-\gamma}}{\Gamma_p(\alpha)} \int_{O<V<U_1} |V|^{\gamma}|U_1 - V|^{\alpha-\frac{p+1}{2}} f(V)\mathrm{d}V = K_{1,U_1,\gamma}^{-\alpha} f \tag{3.62}$$

which is Erdélyi-Kober fractional integral of the first kind of order α and parameter γ in the real matrix-variate case for $\Re(\alpha) > \frac{p-1}{2}$.

Case (2): Let $\phi_1(X_1) = |X_1|^{-\alpha-\frac{p+1}{2}}, \phi_2(X_2) = X_2^{\alpha}$. Then $g_1(U_1)$ of (3.61) becomes the following:

$$\frac{1}{\Gamma_p(\alpha)} \int_{O<V<U_1} |U_1 - V|^{\alpha-\frac{p+1}{2}} f(V)\mathrm{d}V = W_{1,U_1}^{-\alpha} f \tag{3.63}$$

which is Weyl fractional integral of order α and of the first kind in the real matrix-variate case. If V is bounded below by a positive definite constant matrix $A > O$ then it is the Riemann-Liouville fractional integral of the first kind and of order α in the real matrix-variate case for $\Re(\alpha) > \frac{p-1}{2}$. By specializing ϕ_1 and ϕ_2 for $p = 1$ one can derive all first kind fractional integrals of order α in the real scalar case and hence the special cases of (3.61) will give the real matrix-variate versions of all those special cases. In the real scalar case one can also take $x_1^{\delta}, \Re(\delta) > 0$ in the type-1 beta function part in $f_1(x_1)$, instead of x_1, but in the real matrix-variate case for $\delta \neq 1$ there is problem with the Jacobian and hence $\delta = 1$ is to be taken.

References

1. T.W. Anderson, *An Introduction to Multivariate Statistical Analysis* (Wiley, New York, 1971)
2. A.M. Kshirsagar, *Multivariate Analysis* (Marcel Dekker, New York, 1972)
3. A.M. Mathai, *Jacobians of Matrix Transformations and Functions of Matrix Argument* (World Scientific Publishing, New York, 1997)
4. A.M. Mathai, A pathway to matrix-variate gamma and normal densities. Linear Algebra Appl. **396**, 317–328 (2005)
5. A.M. Mathai, Random volumes under a general matrix-variate model. Linear Algebra Appl. **425**, 162–170 (2007)
6. A.M. Mathai, Fractional integrals in the matrix-variate case and connection to statistical distributions. Integral Transforms Spec. Funct. **20**(12), 871–882 (2009)

7. A.M. Mathai, Some properties of Mittag-Leffler functions and matrix-variate analogues: a statistical perspective. Fract. Calc. Appl. Anal. **13**(1), 113–132 (2010)
8. A.M. Mathai, Explicit evaluations of gamma and beta integrals. J. Indian Math. Soc. **81**(1–3), 259–271 (2014)
9. A.M. Mathai, Evaluation of matrix-variate gamma and beta integrals. Appl. Math. Comput. **247**, 312–318 (2014)
10. A.M. Mathai, H.J. Haubold, *Special Functions for Applied Scientists* (Springer, New York, 2008)
11. A.M. Mathai, H.J. Haubold, Matrix-variate statistical distributions and fractional calculus. Fract. Calc. Appl. Anal. **14**(1), 138–155 (2011)
12. A.M. Mathai, H.J. Haubold, Fractional operators in the matrix-variate case. Fract. Calc. Appl. Anal. **16**(2), 469–478 (2013)
13. A.M. Mathai, T. Princy, Multivariate and matrix-variate analogues of Maxwell-Boltzmann and Raleigh densities. Physica A **468**, 668–676 (2017)
14. A.M. Mathai, T. Princy, Analogues of reliability analysis for matrix-variate cases. Linear Algebra Appl. **532**, 287–311 (2017a)
15. A.M. Mathai, S.B. Provost, T. Hayakawa, *Bilinear Forms and Zonal Polynomials*. Lecture Notes in Statistics, vol. 102 (Springer, New York, 1995)
16. S. Thomas, A.M. Mathai, p-content of p-parallelotope and its connection to likelihood ratio statistic. Sankhya Series A **71**(1), 49–63 (2009)
17. M.S. Srivastava, C.G. Khatri, *An Introduction to Multivariate Statistics* (North Holland, New York, 1979)

Chapter 4
Erdélyi-Kober Fractional Integrals in the Many Real Scalar Variables Case

When going from a one-variable function to many-variable function there is no unique one to one correspondence. Many types of multivariate functions can be considered when one has the preselected one-variable function. Hence there is nothing called the multivariate analogue of a univariate operator or univariate integral. Hence we construct one multivariate operator here which is analogous to a one variable Erdélyi-Kober fractional integral operator of the second kind or first kind. Other such analogues can be defined. The second kind fractional integrals will be considered first. In this chapter, multivariate case means the case of many real scalar variables.

Definition 4.1 (Erdélyi-Kober fractional integral of the second kind in the multivariate case) This will be defined as the following fractional integral and denoted as follows:

$$K_{2,u_j,\zeta_j,j=1,\ldots,k}^{-\alpha_j,j=1,\ldots,k}f=\{\prod_{j=1}^{k}\frac{u_j^{\zeta_j}}{\Gamma(\alpha_j)}\}\{\prod_{j=1}^{k}\int_{v_j=u_j}^{\infty}(v_j-u_j)^{\alpha_j-1}v_j^{-\zeta_j-\alpha_j}\}f(v_1,\ldots,v_k)\mathrm{d}v_1\wedge\ldots\wedge\mathrm{d}v_k. \tag{4.1}$$

This definition is parallel to the one in the one variable case. We will now look at various connections to different problems. First we will establish a number of results in connection with statistical distribution theory. We will show that (4.1) can be treated as a constant multiple of a joint density of a number of random variables $u_1,\ldots,u_k$ appearing in different contexts. This type of interpretations for fractional integrals are easier to comprehend.

A. M. Mathai, H. J. Haubold, *Erdélyi–Kober Fractional Calculus*, SpringerBriefs in Mathematical Physics 31, https://doi.org/10.1007/978-981-13-1159-8_4

4.1 Erdélyi-Kober Fractional Integrals of the Second Kind in Multivariate Case as Statistical Densities

Let $x_1, x_2, \ldots, x_k$ be independently distributed type-1 beta random variables with parameters $(\zeta_j + 1, \alpha_j), j = 1, \ldots, k, \zeta_j > -1, \alpha_j > 0, j = 1, \ldots, k$. Usually the parameters in a statistical density are real but the following integrals also exist for complex parameters and in that case the conditions will be $\Re(\zeta) > -1$ and $\Re(\alpha_j) > 0, j = 1, \ldots, k$. That is, the density of x_j is of the form

$$f_j(x_j) = \frac{\Gamma(\alpha_j + \zeta_j + 1)}{\Gamma(\alpha_j)\Gamma(\zeta_j + 1)} x_j^{\zeta_j}(1 - x_j)^{\alpha_j - 1}, 0 < x_j < 1 \tag{4.2}$$

for $\alpha_j > 0, \zeta_j > -1$ and $f_j(x_j) = 0$ elsewhere, $j = 1, \ldots, k$ so that the joint density of $x_1, \ldots, x_k$ is the product $f_1(x_1)\ldots f_k(x_k)$. Let $v_1, \ldots, v_k$ be another sequence of real scalar positive random variables having an arbitrary joint density $f(v_1, \ldots, v_k)$, arbitrary in the sense, any real-valued function f such that $f \geq 0$ for all $v_1, \ldots, v_k$ and $\int_{v_1} \cdots \int_{v_k} f(v_1, \ldots, v_k)\mathrm{d}v_1 \wedge \ldots \wedge \mathrm{d}v_k = 1$. Let the two sets $(x_1, \ldots, x_k)$ and $(v_1, \ldots, v_k)$ be independently distributed in the sense that the joint density of the sets $\{x_1, \ldots, x_k\}$ and $\{v_1, \ldots, v_k\}$ is the product of the joint density of $\{x_1, \ldots, x_k\}$ and the joint density of $\{v_1, \ldots, v_k\}$. Note that the joint density of the set $\{x_1, \ldots, x_k\}$ is $f_1(x_1)\ldots f_k(x_k)$ and the joint density of $\{v_1, \ldots, v_k\}$ is $f(v_1, \ldots, v_k)$. Thus the joint density of both the sets is $f(v_1, \ldots, v_k)\prod_{j=1}^k f_j(x_j)$. Consider the transformation $u_j = x_j v_j$ or $x_j = \frac{u_j}{v_j}, v_j = v_j$ and the Jacobian of the transformation is

$$\{\prod_{j=1}^k \wedge \mathrm{d}x_j\} \wedge \{\prod_{j=1}^k \wedge \mathrm{d}v_j\} = \{\prod_{j=1}^k (v_j)^{-1}\}\{\prod_{j=1}^k \wedge \mathrm{d}u_j\} \wedge \{\prod_{j=1}^k \wedge \mathrm{d}v_j\} \tag{4.3}$$

Then the joint density of $u_1, \ldots, u_k$, denoted by $g_2(u_1, \ldots, u_k)$, is given by

$$\begin{aligned} g_2(u_1, \ldots, u_k) &= \{\prod_{j=1}^k \frac{\Gamma(\alpha_j + \zeta_j + 1)}{\Gamma(\zeta_j + 1)\Gamma(\alpha_j)}\}\{\prod_{j=1}^k \int_{v_j} (\frac{u_j}{v_j})^{\zeta_j}(1 - \frac{u_j}{v_j})^{\alpha_j - 1}\} \\ &\quad \times f(v_1, \ldots, v_k)\mathrm{d}v_1 \wedge \ldots \wedge \mathrm{d}v_k \\ &= \{\prod_{j=1}^k \frac{\Gamma(\alpha_j + \zeta_j + 1)}{\Gamma(\zeta_j + 1)\Gamma(\alpha_j)}\}\{\prod_{j=1}^k u_j^{\zeta_j} \int_{v_j = u_j}^{\infty} (v_j - u_j)^{\alpha_j - 1} v_j^{-\zeta_j - \alpha_j}\} \\ &\quad \times f(v_1, \ldots, v_k)\mathrm{d}v_1 \wedge \ldots \wedge \mathrm{d}v_k. \end{aligned} \tag{4.4}$$

We will use g_2 to denote second kind fractional integrals and g_1 to denote first kind fractional integrals. Therefore one can write

Theorem 4.1 *Let $x_j, u_j, v_j, j = 1, \ldots, k$ be as defined above where $x_1, \ldots, x_k$ are independently type-1 beta distributed with parameters $(\zeta_j + 1, \alpha_j), j = 1, \ldots, k$, $v_1, \ldots, v_k$ having a joint arbitrary density $f(v_1, \ldots, v_k)$ with $(x_1, \ldots, x_k)$ and $(v_1, \ldots, v_k)$ being independently distributed. If the joint density of $u_1, \ldots, u_k$ is denoted as $g_2(u_1, \ldots, u_k)$ then*

$$\{\prod_{j=1}^{k} \frac{\Gamma(\zeta + 1)}{\Gamma(\alpha_j + \zeta_j + 1)}\} g_2(u_1, \ldots, u_k) = K_{2,u_j,\zeta_j,j=1,\ldots,k}^{-\alpha_j,j=1,\ldots,k} f \tag{4.5}$$

for $\Re(\zeta_j) > -1, \Re(\alpha_j) > 0, j = 1, \ldots, k$, where $K_{2,u_j,\zeta_j,j=1,\ldots,k}^{-\alpha_j,j=1,\ldots,k}$ will be called Erdélyi-Kober fractional integral operator of the second kind for the multivariate or for the many real scalar variables case.

In this case we have $x_1, \ldots, x_k$ mutually independently distributed and thus there are a total of $k + 1$ densities involved, the k of $x_1, \ldots, x_k$ and the one of $(v_1, \ldots, v_k)$. Let us see what happens if $x_1, \ldots, x_k$ are not independently distributed but they have a joint density $f_1(x_1, \ldots, x_k)$ and $(v_1, \ldots, v_k)$ having a joint density $f_2(v_1, \ldots, v_k)$. Then we can show that if f_1 can be eventually reduced to independent type-1 beta form, still we can consider Erdélyi-Kober fractional integrals of the second kind in the multivariate case as constant multiples of statistical densities.

Let $(x_1, \ldots, x_k)$ have a joint type-1 Dirichlet density with parameters $(\alpha_1 + 1, \ldots, \alpha_k + 1; \alpha_{k+1}), \Re(\alpha_j) > -1, j = 1, \ldots, k, \Re(\alpha_{k+1}) > 0$, that is,

$$f_1(x_1, \ldots, x_k) = \frac{\Gamma(\alpha_1 + \ldots + \alpha_{k+1} + k)}{\{\prod_{j=1}^{k} \Gamma(\alpha_j + 1)\}\Gamma(\alpha_{k+1})} x_1^{\alpha_1} \ldots x_k^{\alpha_k} (1 - x_1 - \ldots - x_k)^{\alpha_{k+1}-1},$$

$$0 < x_j < 1, j = 1, \ldots, k, 0 < x_1 + \ldots + x_k < 1 \tag{4.6}$$

and $f_1(x_1, \ldots, x_k) = 0$ elsewhere. Let us consider the transformations $x_1 = y_1, x_2 = y_2(1 - y_1), \ldots x_k = (1 - y_1) \ldots (1 - y_{k-1})$ or

$$x_j = y_j(1 - y_1)(1 - y_2) \ldots (1 - y_{j-1}), j = 1, \ldots, k \text{ or}$$

$$y_j = \frac{x_j}{1 - x_1 - \ldots - x_{j-1}}, j = 1, \ldots, k. \tag{4.7}$$

Under this transformation the Jacobian is $(1 - y_1)^{k-1} \ldots (1 - y_{k-1})$. It is easy to show that under this transformation $y_1, \ldots, y_k$ will be independently distributed as type-1 beta variables with the parameters $(\alpha_j + 1, \beta_j)$ with $\beta_j = \alpha_{j+1} + \alpha_{j+2} + \ldots + \alpha_k + (k - j) + \alpha_{k+1}$ or y_j has the density

$$f_j(y_j) = \frac{\Gamma(\alpha_j + 1 + \beta_j)}{\Gamma(\alpha_j + 1)\Gamma(\beta_j)} y_j^{\alpha_j} (1 - y_j)^{\beta_j - 1}, 0 < y_j < 1$$

and $f_j(y_j) = 0$ elsewhere, $\alpha_j > -1, \beta_j > 0, j = 1, \ldots, k$. In the light of these observations, let us consider two sets of positive random variables $(x_1, \ldots, x_k)$ and $(v_1, \ldots, v_k)$ where the two sets are independently distributed with $(x_1, \ldots, x_k)$ having a type-1 Dirichlet distribution. Let $u_j = y_j v_j = v_j(\frac{x_j}{1-x_1-\ldots-x_{j-1}}), j = 1, \ldots, k$. Then following through the same procedure as above we have the following theorem.

Theorem 4.2 *Let $(x_1, \ldots, x_k)$ and $(v_1, \ldots, v_k)$ be two sets of real scalar positive random variables where the two sets are independently distributed. Let $(v_1, \ldots, v_k)$ have a joint density $f_2(v_1, \ldots, v_k) = f(v_1, \ldots, v_k)$ where f is arbitrary, and let $(x_1, \ldots, x_k)$ have a type-1 Dirichlet density with the parameters $(\alpha_1 + 1, \ldots, \alpha_k + 1; \alpha_{k+1})$ or with the density*

$$f_1(x_1, \ldots, x_k) = C_1 \, x_1^{\alpha_1} \ldots x_k^{\alpha_k} (1 - x_1 - \ldots - x_k)^{\alpha_{k+1}-1} \tag{4.8}$$

for $0 < x_j < 1, 0 < x_1 + \ldots + x_k < 1, j = 1, \ldots, k$ and $f_1(x_1, \ldots, x_k) = 0$ elsewhere, where C_1 is the normalizing constant. Let $u_j = v_j(\frac{x_j}{1-x_1-\ldots-x_{j-1}}), j = 1, \ldots, k$. If the joint density of $u_1, \ldots, u_k$ is denoted by $g_{21}(u_1, \ldots, u_k)$ then

$$\{\prod_{j=1}^{k} \frac{\Gamma(\alpha_j + 1)}{\Gamma(\alpha_j + \beta_j + 1)}\} g_{21}(u_1, \ldots, u_k) = K_{2,u_j,\alpha_j,j=1,\ldots,k}^{-\beta_j,j=1,\ldots,k} f \tag{4.9}$$

where $\beta_j = \alpha_{j+1} + \alpha_{j+2} + \ldots + \alpha_k + (k - j) + \alpha_{k+1}, j = 1, \ldots, k$, $\Re(\alpha_j) > -1, j = 1, \ldots, k, \Re(\alpha_{k+1}) > 0, \Re(\beta_j) > 0, j = 1, \ldots, k$.

The above structure indicates that we can consider any multivariate density $f_1(x_1, \ldots, x_k)$ for a set of real scalar positive random variables $(x_1, \ldots, x_k)$ and if we can find a suitable transformation to bring the joint density of the new variables as products of type-1 beta densities then the Erdélyi-Kober fractional integrals of the second kind for the multivariate case can be written in terms of a statistical density as shown above. There are many densities where a transformation can bring $f_1(x_1, \ldots, x_k)$ to product of type-1 beta densities. There are several generalizations of type-1 and type-2 Dirichlet models where suitable transformations exist which can bring a set of mutually independently distributed type-1 beta random variables. We will list one more example of this type before quitting this section.

Let us consider a generalized type-1 Dirichlet model of the following type. Several types of generalizations of the following category are available.

$$\begin{aligned} f_3(x_1, \ldots, x_k) &= C_3 \, x_1^{\alpha_1} (1 - x_1)^{\beta_1} x_2^{\alpha_2} (1 - x_1 - x_2)^{\beta_2} \ldots \\ &\quad \times x_k^{\alpha_k} (1 - x_1 - \ldots - x_k)^{\beta_k + \alpha_{k+1} - 1}, \\ &\quad 0 < x_1 + \ldots + x_j < 1, \, j = 1, \ldots, k \end{aligned} \tag{4.10}$$

and $f_3(x_1, \ldots, x_k) = 0$ elsewhere, where C_3 is a normalizing constant. Let the corresponding density for $(v_1, \ldots, v_k)$ be $f_4(v_1, \ldots, v_k) = f(v_1, \ldots, v_k)$ where f is arbitrary. Let us consider the same transformation as in (4.7). Then we can show that $y_1, \ldots, y_k$ will be mutually independently distributed as type-1 beta random variables with the parameters $(\alpha_j + 1, \gamma_j)$, $j = 1, \ldots, k$ where

$$\gamma_j = \alpha_{j+1} + \alpha_{j+2} + \ldots + \alpha_{k+1} + \beta_j + \beta_{j+1} + \ldots + \beta_k + (k - j) \tag{4.11}$$

for $j = 1, \ldots, k$ with $\Re(\alpha_j) > -1$, $j = 1, \ldots, k$, $\Re(\alpha_{k+1}) > 0$ and $\Re(\gamma_j) > 0$, $j = 1, \ldots, k$.

Theorem 4.3 *Let $x_1, \ldots, x_k$ have a joint density of the form in* (4.10). *Let $v_1, \ldots, v_k$ be another set of real scalar positive random variables having an arbitrary density $f(v_1, \ldots, v_k)$. Between sets let $(x_1, \ldots, x_k)$ and $(v_1, \ldots, v_k)$ be independently distributed. Consider the transformation as in* (4.7) *where*

$$u_j = v_j\left(\frac{x_j}{1 - x_1 - \ldots - x_{j-1}}\right), \; j = 1, \ldots, k.$$

Let the joint density of $u_1, \ldots, u_k$ be denoted by $g_{22}(u_1, \ldots, u_k)$. Then

$$\left\{\prod \frac{\Gamma(\alpha_j + 1)}{\Gamma(\alpha_j + \gamma_j + 1)}\right\} g_{22}(u_1, \ldots, u_k) = K_{2,u_j,\alpha_j,j=1,\ldots k}^{-\gamma_j,j=1,\ldots,k} f \tag{4.12}$$

where $\gamma_j = \alpha_{j+1} + \alpha_{j+2} + \ldots + \alpha_{k+1} + \beta_j + \beta_{j+1} + \ldots + \beta_k + (k - j)$, $j = 1, \ldots, k$ for $\Re(\alpha_j) > -1$, $\Re(\gamma_j) > 0$, $j = 1, \ldots, k$, $\Re(\alpha_{k+1}) > 0$.

4.2 A Pathway Generalization of Erdélyi-Kober Fractional Integral Operator of the Second Kind in the Multivariate Case

Let $x_1, \ldots, x_k$ be independently distributed with x_j having a pathway density given by

$$f_{jp}(x_j) = c_{jp}\, x_j^{\zeta_j} [1 - a_j(1 - q_j)x_j]^{\frac{\eta_j}{1-q_j}} \tag{4.13}$$

for $1 - a_j(1 - q_j)x_j > 0, a_j > 0, q_j < 1, \eta_j > 0, \zeta_j > -1$ and $f_{jp}(x_j) = 0$ elsewhere, where

$$c_{jp} = \frac{[a_j(1 - q_j)]^{\zeta_j+1} \Gamma(\zeta_j + 1 + \frac{\eta_j}{1-q_j} + 1)}{\Gamma(\zeta_j + 1)\Gamma(\frac{\eta_j}{1-q_j} + 1)}. \tag{4.14}$$

Let $v_1, \ldots, v_k$ be real scalar positive random variables with a joint density $f(v_1, \ldots, v_k)$. Let $(x_1, \ldots, x_k)$ and $(v_1, \ldots, v_k)$ be statistically independently distributed. Let $u_j = x_j v_j$, $x_j = \frac{u_j}{v_j}$, $j = 1, \ldots, k$. Then the Jacobian of the transformation is $(v_1 \ldots v_k)^{-1}$. Let the joint density of $u_1, \ldots, u_k$ be denoted by $g_{23}(u_1, \ldots, u_k)$. Then from the standard technique of transformation of variables the density g_{23} is given by

$$g_{23}(u_1, \ldots, u_k) = \{\prod_{j=1}^{k} c_{jp} u_j^{\zeta_j} \int_{v_j = a_j(1-q_j)u_j}^{\infty} v_j^{-\zeta_j - (\frac{\eta_j}{1-q_j}+1)}$$

$$\times [v_j - a(1-q_j)u_j]^{\frac{\eta_j}{1-q_j}}\} f(v_1, \ldots, v_k) \mathrm{d}v_1 \wedge \ldots \wedge \mathrm{d}v_k. \quad (4.15)$$

Hence we may define a pathway extension of Erdélyi-Kober fractional integral of the second kind in the multivariate case.

Definition 4.2 (A Pathway Erdélyi-Kober Fractional Integral of the Second Kind for the Multivariate Case) It will be defined and denoted as follows:

$$K_{2,u_j,\zeta_j,a_j,q_j,j=1,\ldots,k}^{-(\frac{\eta_j}{1-q_j}+1),j=1,\ldots,k} f = \{\prod_{j=1}^{k} \frac{[a_j(1-q_j)]^{\zeta_j+1} u_j^{\zeta_j}}{\Gamma(\frac{\eta_j}{1-q_j}+1)} \int_{v_j > a_j(1-q_j)u_j > 0} v_j^{-\zeta_j - (\frac{\eta_j}{1-q_j}+1)}$$

$$\times (v_j - a_j(1-q_j)u_j)^{\frac{\eta_j}{1-q_j}}\} f(v_1, \ldots, v_k) \mathrm{d}v_1 \wedge \ldots \wedge \mathrm{d}v_k. \quad (4.16)$$

for $a_j > 0, q_j < 1, \eta_j > 0, \Re(\zeta_j) > -1$.

Theorem 4.4 *Let* $x_1, \ldots, x_k$, $v_1, \ldots, v_k$, u_j, $j = 1, \ldots, k$ *and* $g_{23}(u_1, \ldots, u_k)$ *be as defined in* (4.15). *Let the pathway extended Erdélyi-Kober fractional integral of the second kind be as defined in* (4.16). *Then*

$$\{\prod_{j=1}^{k} \frac{\Gamma(\zeta_j+1)}{\Gamma(\zeta_j + \frac{\eta_j}{1-q_j} + 2)}\} g_{23}(u_1, \ldots, u_k) = K_{2,u_j,\zeta_j,a_j,q_j,j=1,\ldots,k}^{-(\frac{\eta_j}{1-q_j}+1),j=1,\ldots,k} f. \quad (4.17)$$

When any particular $q_r \to 1_-$ then we can see the corresponding factor going to the exponential form.

$$\lim_{q_r \to 1_-} \frac{(1-q_r)^{\zeta_r+1} \Gamma(\zeta_r + \frac{\eta_r}{1-q_r} + 2)}{\Gamma(\frac{\eta_r}{1-q_r}+1)} (\frac{u_r}{v_r})^{\zeta_r} \frac{1}{v_r} [1 - a_r(1-q_r)(\frac{u_r}{v_r})]^{\frac{\eta_r}{1-q_r}}$$

$$= \eta_r^{\zeta_r+1} (\frac{u_r}{v_r})^{\zeta_r} \frac{1}{v_r} \mathrm{e}^{-a_r \eta_r (\frac{u_r}{v_r})}, 0 < v_r < \infty. \quad (4.18)$$

Thus, individual q_j's can go to 1 and the corresponding factor will go to exponential form or the correspondingly we get a gamma density structure for that factor.

4.3 Mellin Transform in the Multivariate Case for Erdélyi-Kober Fractional Integral of the Second Kind

The Mellin transform in the multivariate case is defined as

$$M\{f(x_1,\ldots,x_k);s_1,\ldots,s_k\}=\int_0^\infty\ldots\int_0^\infty x_1^{s_1-1}\ldots x_k^{s_k-1}f(x_1,\ldots,x_k)\mathrm{d}x_1\wedge\ldots\wedge\mathrm{d}x_k \tag{4.19}$$

whenever it exists, where $s_1,\ldots,s_k$ in general are complex parameters. Hence for the Erdélyi-Kober fractional integral operator of the second kind we have

$$\begin{aligned}M\{K_{u_j,\zeta_j,j=1,\ldots,k}^{-\alpha_j,j=1,\ldots,k}f(u_1,\ldots,u_k);s_1,\ldots,s_k\}&=\int_0^\infty\ldots\int_0^\infty u_1^{s_1-1}\ldots u_k^{s_k-1}\\&\times\{\prod_{j=1}^k\frac{u_j^{\zeta_j}}{\Gamma(\alpha_j)}\int_{v_j>u_j>0}(v_j-u_j)^{\alpha_j-1}v_j^{-\zeta-\alpha_j}\}\\&\times f(v_1,\ldots,v_k)\mathrm{d}V\wedge\mathrm{d}U\end{aligned} \tag{i}$$

where, for example, $\mathrm{d}U=\mathrm{d}u_1\wedge\ldots\wedge\mathrm{d}u_k$, $\mathrm{d}V=\mathrm{d}v_1\wedge\ldots\wedge\mathrm{d}v_k$. Then the right side of (i) is

$$=\int_0^\infty\ldots\int_0^\infty f(v_1,\ldots,v_k)\{\prod_{j=1}^k v_j^{-\zeta_j-\alpha_j}\}[\{\prod_{j=1}^k\int_0^{v_j}u_j^{s_j+\zeta_j-1}(v_j-u_j)^{\alpha_j-1}\mathrm{d}u_j\}]\mathrm{d}V.$$

Take out v_j, put $y_j=\frac{u_j}{v_j}$ then the integral over u_j will go to

$$v_j^{\zeta_j+\alpha_j+s_j-1}\frac{\Gamma(\alpha_j)\Gamma(\zeta_j+s_j)}{\Gamma(\alpha_j+\zeta_j+s_j)}.$$

Hence the required Mellin transform is

$$\left[\prod_{j=1}^k\frac{\Gamma(\zeta_j+s_j)}{\Gamma(\alpha_j+\zeta_j+s_j)}\right]f^*(s_1,\ldots,s_k)$$

where f^* is the Mellin transform of f. Then we have the following theorem.

Theorem 4.5 *For the Erdélyi-Kober fractional integral of the second kind in the multivariate case as defined in* (4.1) *the Mellin transform, with Mellin parameters* $s_1, \ldots, s_k$ *is given by*

$$M\{K_{2,u_j,\zeta_j,j=1,\ldots,k}^{-\alpha_j,j=1,\ldots,k} f; s_1,\ldots,s_k\} = \left[\prod_{j=1}^{k} \frac{\Gamma(\zeta_j + s_j)}{\Gamma(\alpha_j + \zeta_j + s_j)}\right] f^*(s_1,\ldots,s_k) \tag{4.20}$$

for $\Re(\alpha_j) > 0, \Re(\zeta_j + s_j) > 0, j = 1,\ldots,k$, *where* f^* *is the Mellin transform of* f.

4.4 Erdélyi-Kober Fractional Integral of the First Kind for Multivariate Case

For the first kind integrals the orders $\alpha_1, \ldots, \alpha_k$ will be written as $(-\alpha_j, j = 1,\ldots,k)$ The letter K will stand for Erdélyi-Kober fractional integral. The first kind integrals will be denoted by 1 and the second kind by 2. The orders are written as superscript to K as $-\alpha_j, j = 1,\ldots,k$ and the additional parameters, variables and the kind number will be written as subscripts to K. For example, $K_{1,u_j,\zeta_j,j=1,\ldots,k}^{-\alpha_j,j=1,\ldots,k} f$ will represent Erdélyi-Kober fractional integral of the first kind in the multivariate case of orders $\alpha_j, j = 1,\ldots,k$ and parameters $\zeta_j, j = 1,\ldots,k$.

In the multivariate case we will start with the following definition and notation.

Definition 4.3 (Erdélyi-Kober Fractional Integral of the First Kind in the Multivariate Case)

$$K_{1,u_j,\zeta_j,j=1,\ldots,k}^{-\alpha_j,j=1,\ldots,k} f = \{\prod_{j=1}^{k} \frac{u_j^{-\zeta_j-\alpha_j}}{\Gamma(\alpha_j)} \int_{v_j=0}^{u_j} (u_j - v_j)^{\alpha_j-1} v_j^{\zeta_j}\} f(v_1,\ldots,v_k) \mathrm{d}v_1 \wedge \ldots \wedge \mathrm{d}v_k. \tag{4.21}$$

For $k = 1$, (4.21) corresponds to Erdélyi-Kober fractional integral of the first kind in the real scalar variable case. First, we will derive the integral in (4.21) as a constant multiple of a statistical density. To this end, let $x_1, \ldots, x_k$ be independently distributed type-1 beta random variables with the parameters $(\zeta_j, \alpha_j), j = 1,\ldots,k$ or with the density

$$f_{5j}(x_j) = \frac{\Gamma(\zeta_j + \alpha_j)}{\Gamma(\zeta_j)\Gamma(\alpha_j)} x_j^{\zeta_j-1}(1 - x_j)^{\alpha_j-1}, 0 < x_j < 1, \tag{4.22}$$

for $\alpha_j > 0, \zeta_j > 0$ or $\Re(\alpha_j) > 0, \Re(\zeta_j) > 0$ when the parameters are in the complex domain, and $f_{5j}(x_j) = 0$ otherwise. Let $(v_1, \ldots, v_k)$ be real scalar positive random variables having a joint density $f_6(v_1,\ldots,v_k)=f(v_1,\ldots,v_k)$ where f is arbitrary. Let $u_j=\frac{v_j}{x_j}$, $j=1,\ldots,k$. The Jacobian is given by

$$\mathrm{d}X\wedge\mathrm{d}V=[\prod_{j=1}^{k}(-\frac{v_j}{u_j^2})]\mathrm{d}U\wedge\mathrm{d}V \tag{4.23}$$

where the earlier simplified notation is used. The joint density of $u_1,\ldots,u_k$, following through the earlier steps, denoted by $g_1(u_1,\ldots,u_k)$, is given by

$$g_1(u_1,\ldots,u_k)=\{\prod_{j=1}^{k}\frac{\Gamma(\zeta_j+\alpha_j)}{\Gamma(\zeta_j)\Gamma(\alpha_j)}\}\{\prod_{j=1}^{k}u_j^{-\zeta_j-\alpha_j}\int_{v=0}^{u_j}(u_j-v_j)^{\alpha_j-1}v_j^{\zeta_j}\}f(v_1,\ldots,v_k)\mathrm{d}v_1\wedge\ldots\wedge\mathrm{d}v_k. \tag{4.24}$$

Here the 1 in g_1 stands for the first kind fractional integral. If the j-th set of $(x_1,\ldots,x_k)$ and $(v_1,\ldots,v_k)$ are being considered then the corresponding g_1 will be denoted as g_{1j}. We have the following theorem.

Theorem 4.6 *Let $x_1,\ldots,x_k$ be independently distributed type-1 beta random variables with the parameters (ζ_j,α_j), $j=1,\ldots,k$, and let $v_1,\ldots,v_k$, $u_1,\ldots,u_k$ be as defined in* (4.22) *and* (4.23). *Let the joint density of $u_1,\ldots,u_k$ be denoted by $g_1(u_1,\ldots,u_k)$. Then*

$$\{\prod_{j=1}^{k}\frac{\Gamma(\zeta_j)}{\Gamma(\zeta_j+\alpha_j)}\}g_1(u_1,\ldots,u_k)=K_{1,u_j,\zeta_j,j=1,\ldots,k}^{-\alpha_j,j=1,\ldots,k}f. \tag{4.25}$$

We can have pathway extension to Erdélyi-Kober fractional integral of the first kind, parallel to the results for the case of second kind. Other properties follow parallel to those for the case of the fractional integral of the second kind. We will evaluate the multivariate Mellin transform following through steps parallel to those in the case of the second kind and hence we give the result here as a theorem.

Theorem 4.7 *The Mellin transform, with Mellin parameters $s_1,\ldots,s_k$, for Erdélyi-Kober fractional integral of the first kind in the multivariate case is given by the following:*

$$M\{K_{1,u_j,\zeta_j,j=1,\ldots,k}^{-\alpha_j,j=1,\ldots,k}f;s_1,\ldots,s_k\}=\{\prod_{j=1}^{k}\frac{\Gamma(1+\zeta_j-s)}{\Gamma(1+\alpha_j+\zeta_j-s)}\}f^*(s_1,\ldots,s_k) \tag{4.26}$$

for $\Re(\alpha_j)>0$, $\Re(\zeta_j)>0$, $\Re(s)<1+\Re(\zeta_j)$, $j=1,\ldots,k$ where f^* is the Mellin transform of f.

We can obtain theorems parallel to the ones for the second kind integrals. Some of these will be stated here without proofs. The derivations are parallel to those for the second kind integrals and hence omitted.

Theorem 4.8 *Let $x_1, \ldots, x_k$ be independently distributed as the pathway model in* (4.13) *with ζ_j replaced by $\zeta_j - 1$, $j = 1, \ldots, k$. Let $(v_1, \ldots, v_k)$ have a joint arbitrary density $f(v_1, \ldots, v_k)$. Let $(x_1, \ldots, x_k)$ and $(v_1, \ldots, v_k)$ be independently distributed. Let $u_j = \frac{v_j}{x_j}$ or $x_j = \frac{v_j}{u_j}, j = 1, \ldots, k$. Let the joint density of $u_1, \ldots, u_k$ be again denoted by $g_{11}(u_1, \ldots, u_k)$. Then*

$$K_{1,u_j,\zeta_j,a_j,q_j,j=1,\ldots,k}^{-(\frac{\eta_j}{1-q_j}+1),j=1,\ldots,k} f = \{\prod_{j=1}^{k} \frac{u_j^{-\zeta_j-(\frac{\eta_j}{1-q_j}+1)}[a_j(1-q_j)]^{\zeta_j}}{\Gamma(\frac{\eta_j}{1-q_j}+1)} \int_{v_j=0}^{\frac{u_j}{1-a_j(1-q_j)}} [u_j - a_j(1-q_j)v_j]^{\frac{\eta_j}{1-q_j}} v_j^{\zeta_j}\}$$

$$\times f(v_1, \ldots, v_k) \mathrm{d}v_1 \wedge \ldots \wedge \mathrm{d}v_k$$

$$= \{\prod_{j=1}^{k} \frac{\Gamma(\zeta_j)}{\Gamma(\zeta_j + \frac{\eta_j}{1-q_j} + 1)}\} g_{11}(u_1, \ldots, u_k) \tag{4.27}$$

for $a_j > 0, q_j < 1, \eta_j > 0, \Re(\zeta_j) > 0, j = 1, \ldots, k$.

From here we can have a definition for a pathway extension of Erdélyi-Kober fractional integral of the first kind in the multivariate case.

Definition 4.4 (Erdélyi-Kober fractional integral of the first kind in the multivariate case) Let the variables and parameters be as defined in Theorem 4.8. Then the pathway extended Erdélyi-Kober fractional integral of the first kind in the multivariate case is defined and denoted as follows:

$$K_{1,u_j,\zeta_j,a_j,q_j,j=1,\ldots,k}^{-(\frac{\eta_j}{1-q_j}+1),j=1,\ldots,k} f = \{\prod_{j=1}^{k} \frac{u_j^{-\zeta_j-(\frac{\eta_j}{1-q_j}+1)}[a_j(1-q_j)]^{\zeta_j}}{\Gamma(\frac{\eta_j}{1-q_j}+1)}$$

$$\times \int_{v_j=0}^{\frac{u_j}{1-a_j(1-q_j)}} [u_j - a_j(1-q_j)v_j]^{\frac{\eta_j}{1-q_j}} v_j^{\zeta_j}\}$$

$$\times f(v_1, \ldots, v_k) \mathrm{d}v_1 \wedge \ldots \wedge \mathrm{d}v_k \tag{4.28}$$

for $a_j > 0, q_j < 1, \eta_j > 0, \Re(\zeta_j) > 0, j = 1, \ldots, k$.

We can list several theorems when $x_1, \ldots, x_k$ are not independently distributed. Two such cases will be listed here without proofs. The proofs will be parallel to those in the second kind cases and hence omitted.

Theorem 4.9 *Let $x_1, \ldots, x_k$ have a type-1 Dirichlet density as in* (4.6) *with α_j replaced by $\alpha_j - 1$ for $j = 1, \ldots, k$ and let $(v_1, \ldots, v_k)$ have an arbitrary joint density $f(v_1, \ldots, v_k)$ where $(x_1, \ldots, x_k)$ and $(v_1, \ldots, v_k)$ are independently distributed. Let $y_j = \frac{x_j}{1-x_1-\ldots-x_{j-1}}$ and let $y_j = \frac{v_j}{u_j}, j = 1, \ldots, k$. Let the joint density of $u_1, \ldots, u_k$ be denoted by $g_{12}(u_1, \ldots, u_k)$. Let $\beta_j = \alpha_{j+1} + \alpha_{j+2} + \ldots + \alpha_{k+1}$. Then*

$$K_{1,u_j,\alpha_j,j=1,\ldots,k}^{-\beta_j,j=1,\ldots,k} f = [\prod_{j=1}^{k} \frac{\Gamma(\alpha_j)}{\Gamma(\alpha_j+\beta_j)}] g_{12}(u_1,\ldots,u_k). \tag{4.29}$$

The next theorem is parallel to Theorem 4.3 for the fractional integral of the second kind.

Theorem 4.10 *Let $x_1,\ldots,x_k$ have a joint density as in (4.10) with α_j replaced by $\alpha_j - 1, j = 1,\ldots,k$. Let $y_j = \frac{x_j}{1-x_1-\ldots-x_{j-1}}$ as defined in (4.7). Let $(v_1,\ldots,v_k)$ have an arbitrary joint density $f(v_1,\ldots,v_k)$. Let $(x_1,\ldots,x_k)$ and $(v_1,\ldots,v_k)$ be independently distributed. Let $u_j = \frac{v_j}{y_j}$ or $y_j = \frac{v_j}{u_j}, j = 1,\ldots,k$. Let $\gamma_j = \alpha_{j+1} + \alpha_{j+2} + \ldots + \alpha_{k+1} + \beta_j + \beta_{j+1} + \ldots + \beta_k$. Then*

$$K_{1,u_j,\alpha_j,j=1,\ldots,k}^{-\gamma_j,j=1,\ldots,k} f = \{\prod_{j=1}^{k} \frac{\Gamma(\alpha_j)}{\Gamma(\alpha_j+\gamma_j)}\} g_{12}(u_1,\ldots,u_k) \tag{4.30}$$

for $\Re(\alpha_j) > 0, \Re(\gamma_j) > 0, j = 1,\ldots,k$.

4.5 A General Definition for First and Second Kind Fractional Integrals in the Multivariate Case

We recall the general definition for the one matrix-variate case from Sect. 3.8 of Chap. 3. By introducing prefixed functions ϕ_1 and ϕ_2 we can give some general definitions for the multivariate case. Let $\{x_1,\ldots,x_k\}$ and $\{v_1,\ldots,v_k\}$ be the two sets of real scalar positive variables. Let the joint functions associated with these sets be denoted by $f_1(x_1,\ldots,x_k)$ and $f_2(v_1,\ldots,v_k)$ respectively. If x_j's and v_j's are random variables then we may take f_1 and f_2 as the corresponding joint densities. Let f_1 and f_2 be of the following forms:

$$f_1(x_1,\ldots,x_k) = \phi_1(x_1,\ldots,x_k) \prod_{j=1}^{k} \frac{(1-x_j)^{\alpha_j-1}}{\Gamma(\alpha_j)} \quad \text{and}$$

$$f_2(v_1,\ldots,v_k) = \phi_2(v_1,\ldots,v_k) f(v_1,\ldots,v_k) \tag{4.31}$$

where $\Re(\alpha_j) > 0, j = 1,\ldots,k$; ϕ_1 and ϕ_2 are pre-specified functions and $f(v_1,\ldots,v_k)$ is an arbitrary function. Consider the case of second kind integrals first for convenience. Let $u_j = x_j v_j, x_j = \frac{u_j}{v_j}, j = 1,\ldots,k$. Consider the transformation $(x_j, v_j) \to (u_j, v_j), j = 1,\ldots,k$. Then the Jacobian is $(v_1 \ldots v_k)^{-1}$. Let the joint density of these products $u_j = x_j v_j$ be denoted by $g_2(u_1,\ldots,u_k)$. If the variables are random variables and if the corresponding functions are statistical densities then $g_2(u_1,\ldots,u_k)$ is the joint density of $u_1,\ldots,u_k$, otherwise it is the

Mellin convolution of the product. Then

$$g_2(u_1,\ldots,u_k)=\{\prod_{j=1}^{k}\frac{1}{\Gamma(\alpha_j)}\}\int_{v_1}\ldots\int_{v_k}(v_1\ldots v_k)^{-1}\phi_1(\frac{u_1}{v_1},\ldots,\frac{u_k}{v_k})$$
$$\times\,\phi_2(v_1,\ldots,v_k)f(v_1,\ldots,v_k)\{\prod_{j=1}^{k}(1-\frac{u_j}{v_j})^{\alpha_j-1}\}\mathrm{d}v_1\wedge\ldots\wedge\mathrm{d}v_k. \tag{4.32}$$

Definition 4.5 (General definition for fractional integral of the second kind in the multivariate case) The integral on the right side of (4.32) is called fractional integral of the second kind of orders $\alpha_1,\ldots,\alpha_k$ in the multivariate or many real scalar variable case, where ϕ_1 and ϕ_2 are prefixed functions and f is an arbitrary function.

Before we give a general definition for fractional integral of the first kind, let us examine one special case here for the sake of illustration.

4.5.1 A Special Case of (4.32)

Let $\phi_1(x_1,\ldots,x_k)=\prod_{j=1}^{k}x_j^{\zeta_j}$ and $\phi_2=1$. Then (4.32) will reduce to the following, denoted by $g_{21}(u_1,\ldots,u_k)$:

$$g_{21}(u_1,\ldots,u_k)=\{\prod_{j=1}^{k}\frac{u_j^{\zeta_j}}{\Gamma(\alpha_j)}\int_{v_j>u_j>0}v_j^{-\zeta_j-\alpha_j}(v_j-u_j)^{\alpha_j-1}\}$$
$$\times\,f(v_1,\ldots,v_k)\mathrm{d}v_1\wedge\ldots\wedge\mathrm{d}v_k,\ \Re(\alpha_j)>0,\ j=1,\ldots,k. \tag{4.33}$$

Observe that (4.33) is the Erdélyi-Kober fractional integral of the second kind of orders $\alpha_1,\ldots,\alpha_k$ and parameters $\zeta_1,\ldots,\zeta_k$.

Now, let us give a general definition for fractional integrals of the first kind. In order to avoid too many symbols, we will use the same $u_1,\ldots,u_k$ to denote ratios also. Let $u_j=\frac{v_j}{x_j}$, or $x_j=\frac{v_j}{u_j}, v_j=v_j, j=1,\ldots,k$. Then the Jacobian is $\prod_{j=1}^{k}\frac{v_j}{u_j^2}$. Let the joint density of these ratios $u_1,\ldots,u_k$ be denoted by $g_1(u_1,\ldots,u_k)$. Then

$$g_1(u_1,\ldots,u_k)=\{\prod_{j=1}^{k}\frac{1}{\Gamma(\alpha_j)}\}\int_{v_1}\cdots\int_{v_k}\phi_1(\frac{v_1}{u_1},\ldots,\frac{v_k}{u_k})\{\prod_{j=1}^{k}(1-\frac{v_j}{u_j})^{\alpha_j-1}\}$$
$$\times\phi_2(v_1,\ldots,v_k)f(v_1,\ldots,v_k)\{\prod_{j=1}^{k}\frac{v_j}{u_j^2}\}\mathrm{d}v_1\wedge\ldots\wedge\mathrm{d}v_k \tag{4.34}$$

for $\Re(\alpha_j)>0,\ j=1,\ldots,k$.

Definition 4.6 (Fractional Integral of the First Kind in the Multivariate Case) A general definition for fractional integral of the first kind of orders $\alpha_1,\ldots,\alpha_k$ is the right side in (4.34), where ϕ_1 and ϕ_2 are prefixed functions and f is an arbitrary function.

For the sake of illustration we will examine one special case here.

4.5.2 Special Case of (4.34)

Let $\phi_1(x_1,\ldots,x_k) = \prod_{j=1}^{k}x_j^{\zeta_j-1}$ and $\phi_2 = 1$. Then (4.34) reduces to the following, denoted by $g_{11}(u_1,\ldots,u_k)$.

$$g_{11}(u_1,\ldots,u_k)=\{\prod_{j=1}^{k}\frac{u_j^{-\zeta_j-\alpha_j}}{\Gamma(\alpha_j)}\}\{\prod_{j=1}^{k}\int_{0<v_j<u_j}v_j^{\zeta_j}(u_j-v_j)^{\alpha_j-1}\}$$
$$\times f(v_1,\ldots,v_k)\mathrm{d}v_1\wedge\ldots\wedge\mathrm{d}v_k,\ \Re(\alpha_j)>0,\ j=1,\ldots,k. \tag{4.35}$$

Observe that (4.35) is Eerdélyi-Kober fractional integral of the first kind of orders $\alpha_1,\ldots,\alpha_k$ and parameters $\zeta_1,\ldots,\zeta_k$. For more details, see [1].

Reference

1. A.M. Mathai, Fractional integral operators involving many matrix variables. Linear Algebra Appl. **446**, 196–215 (2014)

Chapter 5
Erdélyi-Kober Fractional Integrals Involving Many Real Matrices

All the matrices appearing in this chapter are $p \times p$ real positive definite unless stated otherwise. In order to avoid too many symbols we will use $u_1 = \frac{x_2}{x_1}$ for the ratio of x_2 to x_1 in the real scalar variable case, $U_1 = X_2^{\frac{1}{2}} X_1^{-1} X_2^{\frac{1}{2}}$, symmetric ratio, in the real $p \times p$ matrix-variate case. The corresponding density of u_1 and U_1 will be indicated by g_1; we will use $u_2 = x_1 x_2$ for the product in the real scalar variable case and $U_2 = X_2^{\frac{1}{2}} X_1 X_2^{\frac{1}{2}}$, the symmetric product, in the real $p \times p$ matrix-variate case. The corresponding density of u_2 or U_2 will be indicated by g_2. If x_1 and x_2 are statistically independently distributed real scalar random variables, and X_1 and X_2 are statistically independently distributed real matrix-variate random variables, then $g_2(u_2)$ or $g_2(U_2)$ and $g_1(u_1)$ or $g_1(U_1)$ will denote product and ratio distributions or M-convolutions of product and ratio whatever be the set of variables. In all the preceding chapters the basic claim is that fractional integrals are of two kinds, the first kind or left-sided and the second kind or right-sided. The first kind of fractional integrals belong to the class of Mellin convolution of a ratio and the second kind of fractional integrals belong to the class of Mellin convolution of a product. In the matrix-variate case these will be M-convolutions of ratio and product respectively. If the variables are random variables then g_1 and g_2 correspond to the densities of ratio and product respectively. We will give the following formal definition of fractional integral operators of the first kind and second kind. For the sake of completeness we will recall the general definitions from Chaps. 3 and 4.

Definition 5.1 (Fractional integral of the first kind in one scalar or matrix variable case) A fractional integral of the first kind of order α and of one scalar or matrix variable is a Mellin convolution of a ratio with the first function $f_1(x_1)$ is of the form

$$f_1(x_1) = \frac{\phi_1(x_1)(1 - x_1)^{\alpha-1}}{\Gamma(\alpha)}, \Re(\alpha) > 0 \tag{5.1}$$

A. M. Mathai, H. J. Haubold, *Erdélyi–Kober Fractional Calculus*, SpringerBriefs in Mathematical Physics 31, https://doi.org/10.1007/978-981-13-1159-8_5

in the real scalar variable case,

$$f_1(X_1) = \frac{\phi(X_1)|I - X_1|^{\alpha - \frac{p+1}{2}}}{\Gamma_p(\alpha)}, \Re(\alpha) > \frac{p-1}{2} \tag{5.2}$$

in the single matrix variable case, where $\phi_1(x_1)$ and $\phi_1(X_1)$ are specified functions, and $f_2(x_2) = \phi_2(x_2)f(x_2)$ where $\phi_2(x_2)$ is a specified function and $f(x_2)$ is an arbitrary function, in the real scalar case, and $f_2(X_2) = \phi_2(X_2)f(X_2)$ in the real matrix-variate case where X_2 is a $p \times p$ real positive definite matrix, so that the fractional integral of the first kind of order α is given by the Mellin convolution formula for a ratio, namely

$$g_1(u_1) = \frac{1}{\Gamma(\alpha)} \int_{0<v<u_1} \phi_1(\frac{v}{u_1})(1 - \frac{v}{u_1})^{\alpha-1} \phi_2(v) \frac{v}{u_1^2} f(v)\mathrm{d}v \tag{5.3}$$

in the scalar variable case, and for one matrix variable case

$$g_1(U_1) = \int_{O<V<U_1} \phi_1(V^{\frac{1}{2}}U_1^{-1}V^{\frac{1}{2}}) \frac{1}{\Gamma_p(\alpha)} |I - V^{\frac{1}{2}}U_1^{-1}V^{\frac{1}{2}}|^{\alpha - \frac{p+1}{2}}$$
$$\times |V|^{\frac{p+1}{2}} |U_1|^{-(p+1)} \phi_2(V) f(V)\mathrm{d}V, \Re(\alpha) > \frac{p-1}{2} \tag{5.4}$$

where, for example, $V^{\frac{1}{2}}$ is the real positive definite square root of the real positive definite matrix V.

Definition 5.2 (Fractional integral of the second kind of order α for the one positive real scalar variable case or one $p \times p$ real positive definite matrix-variate case) Let f_1 and f_2 be as given in Definition 5.1. Then the fractional integral of the second kind of order α is defined as the Mellin convolution of a product, denoted by g_2, and given by

$$g_2(u_2) = \frac{1}{\Gamma(\alpha)} \int_{v>u_2} \frac{1}{v} \phi_1(\frac{u_2}{v})(1 - \frac{u_2}{v})^{\alpha-1} \phi_2(v) f(v)\mathrm{d}v, \Re(\alpha) > 0 \tag{5.5}$$

for the real scalar case, and for the real matrix-variate case

$$g_2(U_2) = \frac{1}{\Gamma_p(\alpha)} \int_{V>U_2>O} |V|^{-\frac{p+1}{2}} \phi_1(V^{-\frac{1}{2}}U_2V^{-\frac{1}{2}}) |I - V^{-\frac{1}{2}}U_2V^{-\frac{1}{2}}|^{\alpha - \frac{p+1}{2}}$$
$$\times \phi_2(V) f(V)\mathrm{d}V, \Re(\alpha) > \frac{p-1}{2}. \tag{5.6}$$

Note that if $f_1(x_1)$ and $f_2(x_2)$ are statistical densities of real positive scalar random variables x_1 and x_2 then $g_1(u_1)$ will represent the density of the ratio $u_1 = \frac{x_2}{x_1}$. In the matrix case $g_1(U_1)$ will represent the density of $U_1 = X_2^{\frac{1}{2}} X_1^{-1} X_2^{\frac{1}{2}}$, the symmetric ratio, where $X_1 = V^{\frac{1}{2}}U^{-1}V^{\frac{1}{2}}$ and $X_2 = V$. Similarly, $g_2(u_2)$ and $g_2(U_2)$ will represent the density of the product $u_2 = x_1x_2$ and $U_2 = X_2^{\frac{1}{2}}X_1X_2^{\frac{1}{2}}$ respectively. This is the statistical connection and later we will see that Erdélyi-Kober fractional integrals are constant multiples of statistical densities when f_1 and f_2 are densities. Special cases of specified functions ϕ_1 and ϕ_2 will give various fractional integrals available in the literature for $p = 1$ or for real scalar case. Then the theory of real scalar case can be extended to that of real matrix-variate case. Some of these are given in Chaps. 2 and 3 and hence these will not be repeated here.

Definition 5.3 (Fractional integral of the first kind of orders $\alpha_1, \ldots, \alpha_k$ in the multivariate case) Let $f_1(x_1, \ldots, x_k)$ and $f_2(v_1, \ldots, v_k)$ be real-valued scalar functions of the scalar variables $x_1, \ldots, x_k$ and $v_1, \ldots, v_k$ respectively, where f_1 is of the form

$$f_1 = \phi_1(x_1, \ldots, x_k)\{\prod_{j=1}^{k} \frac{(1-x_j)^{\alpha_j-1}}{\Gamma(\alpha_j)}\}, \Re(\alpha_j) > 0, j = 1, \ldots, k$$

and

$$f_2(v_1, \ldots, v_k) = \phi_2(v_1, \ldots, v_k)f(v_1, \ldots, v_k)$$

where f is an arbitrary function and ϕ_1 and ϕ_2 are prefixed functions. Then the fractional integral of the first kind or left-sided of orders $\alpha_1, \ldots, \alpha_k$ in the multivariate scalar case is given by

$$\begin{aligned} g_1(u_1, \ldots, u_k) &= \{\prod_{j=1}^{k} \frac{1}{\Gamma(\alpha_j)} \int_{v_j<u_j} (1 - \frac{v_j}{u_j})^{\alpha_j-1} \frac{v_j}{u_j^2}\} \\ &\quad \times \phi_1(\frac{v_1}{u_1}, \ldots, \frac{v_k}{u_k})\phi_2(v_1, \ldots, v_k)f(v_1, \ldots, v_k)\mathrm{d}V, \\ \mathrm{d}V &= \mathrm{d}v_1 \wedge \ldots \wedge \mathrm{d}v_k. \end{aligned} \tag{5.7}$$

In the corresponding many matrix-variate case $g_1(U_1, \ldots, U_k)$ is given by the following:

$$\begin{aligned} g_1(U_1, \ldots, U_k) = &\{\prod_{j=1}^{k} \frac{1}{\Gamma_p(\alpha_j)} \int_{O<V_j<U_j} |I - V_j^{\frac{1}{2}}U_j^{-1}V_j^{\frac{1}{2}}|^{\alpha_j - \frac{p+1}{2}} |V_j|^{\frac{p+1}{2}} |U_j|^{-(p+1)}\} \\ &\times \phi_1(V_1^{\frac{1}{2}}U_1^{-1}V_1^{\frac{1}{2}}, \ldots, V_k^{\frac{1}{2}}U_k^{-1}V_k^{\frac{1}{2}}) \\ &\times \phi_2(V_1, \ldots, V_k)f(V_1, \ldots, V_k)\mathrm{d}V, \mathrm{d}V = \mathrm{d}V_1 \wedge \ldots \wedge \mathrm{d}V_k, \end{aligned} \tag{5.8}$$

for $\Re(\alpha_j) > \frac{p-1}{2}, j = 1, \ldots, k$.

Definition 5.4 (Fractional integral of the second kind of orders $\alpha_1,\ldots,\alpha_k$ in the multivariate case) Let $f_1(x_1,\ldots,x_k)$ and $f_2(v_1,\ldots,v_k)$ be real-valued scalar functions of the scalar variables $x_1,\ldots,x_k$ and $v_1,\ldots,v_k$ respectively, where f_1 is of the form

$$f_1=\phi_1(x_1,\ldots,x_k)\{\prod_{j=1}^{k}\frac{(1-x_j)^{\alpha_j-1}}{\Gamma(\alpha_j)}\},\Re(\alpha_j)>0,\ j=1,\ldots,k$$

and $f_2=\phi_2(v_1,\ldots,v_k)f(v_1,\ldots,v_k)$ where f is an arbitrary function and ϕ_1 and ϕ_2 are prefixed functions. Then the fractional integral of the second kind of orders $\alpha_1,\ldots,\alpha_k$ in the multivariate scalar case is given by

$$\begin{aligned}g_2(u_1,\ldots,u_k)&=\{\prod_{j=1}^{k}\frac{1}{\Gamma(\alpha_j)}\int_{v_j>u_j>0}(1-\frac{u_j}{v_j})^{\alpha_j-1}\frac{1}{v_j}\}\\&\quad\times\phi_1(\frac{u_1}{v_1},\ldots,\frac{u_k}{v_k})\phi_2(v_1,\ldots,v_k)f(v_1,\ldots,v_k)\mathrm{d}V,\\ \mathrm{d}V&=\mathrm{d}v_1\wedge\ldots\wedge\mathrm{d}v_k,\end{aligned}\tag{5.9}$$

for $\Re(\alpha_j)>0,\ j=1,\ldots,k$. In the corresponding many matrix-variate case, g_2 is given by

$$\begin{aligned}g_2(U_1,\ldots,U_k)&=\int_{V_1>U_1>O}\cdots\int_{V_k>U_k>O}\phi_1(V_1^{-\frac{1}{2}}U_1V_1^{-\frac{1}{2}},\ldots,V_k^{-\frac{1}{2}}U_kV_k^{-\frac{1}{2}})\\&\quad\times\{\prod_{j=1}^{k}\frac{1}{\Gamma_p(\alpha_j)}|I-V_j^{-\frac{1}{2}}U_jV_j^{-\frac{1}{2}}|^{\alpha_j-\frac{p+1}{2}}|V_j|^{-\frac{p+1}{2}}\}\\&\quad\times\phi_2(V_1,\ldots,V_k)f(V_1,\ldots,V_k)\mathrm{d}V,\ \mathrm{d}V=\mathrm{d}V_1\wedge\ldots\wedge\mathrm{d}V_k,\end{aligned}\tag{5.10}$$

for $\Re(\alpha_j)>\frac{p-1}{2},\ j=1,\ldots,k$.

5.1 Second Kind Fractional Integrals in the Many Matrix-variate Case and Statistical Densities

Let $X_1,\ldots,X_k$ and $V_1,\ldots,V_k$ be two sequences of positive definite $p\times p$ matrix random variables where between the sets the two sets are statistically independently distributed. Further, let $X_1,\ldots,X_k$ be mutually independently distributed type-1 real matrix-variate beta random variables with the parameters $(\zeta_j+\frac{p+1}{2},\alpha_j),\ j=1,\ldots,k$. That is, X_j has the density

$$f_j(X_j) = \frac{\Gamma_p(\alpha_j + \zeta_j + \frac{p+1}{2})}{\Gamma_p(\alpha_j)\Gamma_p(\zeta_j + \frac{p+1}{2})}|X_j|^{\zeta_j}|I - X_j|^{\alpha_j - \frac{p+1}{2}} \tag{5.11}$$

for $O < X_j < I$, $\Re(\alpha_j) > \frac{p-1}{2}, \Re(\zeta_j) > -1$ and $f_j(X_j) = 0, j = 1, \ldots, k$ elsewhere. Let the joint density of $V_1, \ldots, V_k$ be $f_2(V_1, \ldots, V_k) = f(V_1, \ldots, V_k)$ where f is arbitrary. Let $U_j = V_j^{\frac{1}{2}} X_j V_j^{\frac{1}{2}}, j = 1, \ldots, k$. Consider the transformation $X_j = V_j^{-\frac{1}{2}} U_j V_j^{-\frac{1}{2}}, V_j = V_j, j = 1, \ldots, k$ then the Jacobian is $|V_1|^{-\frac{p+1}{2}} \ldots |V_k|^{-\frac{p+1}{2}}$. Substituting in (5.11) the joint density of $U_1, \ldots, U_k$, denoted by $g_2(U_1, \ldots, U_k)$, is given by

$$\begin{aligned}
g_2(U_1, \ldots, U_k) &= \{\prod_{j=1}^{k} \frac{\Gamma_p(\alpha_j + \zeta_j + \frac{p+1}{2})}{\Gamma_p(\alpha_j)\Gamma_p(\zeta_j + \frac{p+1}{2})} \\
&\quad \times \int_{V_j > U_j > O} |V_j|^{-\frac{p+1}{2}} |V_j^{-\frac{1}{2}} U_j V_j^{-\frac{1}{2}}|^{\zeta_j} |I - V^{-\frac{1}{2}} U_j V_j^{-\frac{1}{2}}|^{\alpha_j - \frac{p+1}{2}}\} \\
&\quad \times f(V_1, \ldots, V_k) \mathrm{d}V_1 \wedge \ldots \wedge \mathrm{d}V_k \\
&= \{\prod_{j=1}^{k} \frac{\Gamma_p(\alpha_j + \zeta_j + \frac{p+1}{2})}{\Gamma_p(\alpha_j)\Gamma_p(\zeta_j + \frac{p+1}{2})} \frac{|U_j|^{\zeta_j}}{\Gamma_p(\alpha_j)} \\
&\quad \times \int_{V_j > U_j > O} |V_j|^{-\zeta_j - \alpha_j} |V_j - U_j|^{\alpha_j - \frac{p+1}{2}}\} f(V_1, \ldots, V_k) \mathrm{d}V, \\
\mathrm{d}V &= \mathrm{d}V_1 \wedge \ldots \wedge \mathrm{d}V_k
\end{aligned} \tag{5.12}$$

for $\Re(\alpha_j) > \frac{p-1}{2}, \Re(\zeta_j) > -1, j = 1, \ldots, k$. Hence we will define Erdélyi-Kober fractional integral of the second kind and of orders $(\alpha_1, \ldots, \alpha_k)$ for the many matrix-variate case , and denote as follows:

$$\begin{aligned}
K_{2,U_j,\zeta_j,j=1,\ldots,k}^{-\alpha_j,j=1,\ldots,k} f &= \{\prod_{j=1}^{k} \frac{|U_j|^{\zeta_j}}{\Gamma_p(\alpha_j)} \int_{V_j > U_j > O} |V_j|^{-\zeta_j - \alpha_j} |V_j - U_j|^{\alpha_j - \frac{p+1}{2}}\} \\
&\quad \times f(V_1, \ldots, V_k) \mathrm{d}V_1 \wedge \ldots \wedge \mathrm{d}V_k, \Re(\alpha_j) > \frac{p-1}{2}, j = 1, \ldots, k.
\end{aligned} \tag{5.13}$$

Therefore this Erdélyi-Kober fractional integral is a constant times a statistical density function, namely,

$$\{\prod_{j=1}^{k} \frac{\Gamma_p(\zeta_j + \frac{p+1}{2})}{\Gamma_p(\alpha_j + \zeta_j + \frac{p+1}{2})}\} g_2(U_1, \ldots, U_k) = K_{2,U_j,\zeta_j,j=1,\ldots,k}^{-\alpha_j,j=1,\ldots,k} f \tag{5.14}$$

for $\Re(\alpha_j) > \frac{p-1}{2}, \Re(\zeta_j) > -1, j = 1, \ldots, k$. Now, let us consider f_1 having a joint density of the positive definite matrix variables $X_1, \ldots, X_k$ and f_2 is a joint density of the positive definite matrix variables $V_1, \ldots, V_k$ where the two sets are independently distributed. Some such joint densities may be seen from Kurian et al. [1], Mathai [2], Mathai and Provost [3]. Then we can have several interesting results where the Erdélyi-Kober fractional integral of (5.14) will become constant multiples of statistical densities coming from various considerations.

Theorem 5.1 *Let the two sets $X_1, \ldots, X_k$ and $V_1, \ldots, V_k$ of real positive definite matrix random variables be independently distributed. Further, let $X_1, \ldots, X_k$ have a joint type-1 Dirichlet density with the parameters $(\zeta_j + \frac{p+1}{2}, j = 1, \ldots, k; \zeta_{k+1})$, $j = 1, \ldots, k$. Consider the transformation*

$$X_1 = Y_1$$

$$X_2 = (I - Y_1)^{\frac{1}{2}} Y_2 (I - Y_1)^{\frac{1}{2}}$$

$$X_j = (I - Y_1')^{\frac{1}{2}} \ldots (I - Y_{j-1})^{\frac{1}{2}} Y_j (I - Y_{j-1})^{\frac{1}{2}} \ldots (I - Y_1')^{\frac{1}{2}}, j = 2, \ldots, k. \tag{5.15}$$

$$= B_{j-1} Y_j B_{j-1}', B_{j-1} = (I - Y_1)^{\frac{1}{2}} \ldots (I - Y_{j-1})^{\frac{1}{2}}$$

Or if we write $I - X_1 - \ldots - X_{j-1} = CC'$ then

$$Y_j = C^{-1} X_j (C')^{-1}, j = 2, \ldots, k, Y_1 = X_1. \tag{5.16}$$

Since the matrices are symmetric the transposes are themselves except for the change in the order of a product. What we need is a representation of the form CC'. Consider the transformation

$$U_j = V_j^{\frac{1}{2}} Y_j V_j^{\frac{1}{2}}, \; Y_j = V_j^{-\frac{1}{2}} U_j V_j^{-\frac{1}{2}}, j = 1, \ldots, k. \tag{5.17}$$

Then the joint density of $U_1, \ldots, U_k$ is constant times the generalized Erdélyi-Kober fractional integral of the second kind defined in (5.14).

Proof Under the transformation in (5.15) or (5.16) the Jacobian is

$$J = |I - Y_1|^{(k-1)(\frac{p+1}{2})} |I - Y_2|^{(k-2)(\frac{p+1}{2})} \ldots |I - Y_k|^{\frac{p+1}{2}} \tag{5.18}$$

and that $Y_1, \ldots, Y_k$ are independently distributed as type-1 real matrix-variate beta random variables with the parameters $(\zeta_j + \frac{p+1}{2}, \beta_j), j = 1, \ldots, k$ where

$$\beta_j = \zeta_{j+1} + \zeta_{j+2} + \ldots + \zeta_{k+1} + (k - j)(\frac{p+1}{2}) \tag{5.19}$$

see, for example Mathai [2]. Now, it is equivalent to the situation in (5.13) and (5.14) with Y_j's standing in place of the independently distributed X_j's, that is, $U_j = V_j^{\frac{1}{2}} Y_j V_j^{\frac{1}{2}}$, and hence from (5.14) we have the following result:

$$\{\prod_{j=1}^{k} \frac{\Gamma_p(\zeta_j + \frac{p+1}{2})}{\Gamma_p(\beta_j + \zeta_j + \frac{p+1}{2})}\} g_2(U_1, \ldots, U_k) = K_{2,U_j,\zeta_j,j=1,\ldots,k}^{-\beta_j,j=1,\ldots,k} f \tag{5.20}$$

where β_j is given in (5.19), $\Re(\beta_j) > \frac{p-1}{2}$, $\Re(\zeta_j) > -1$, $j = 1, \ldots, k$, $\Re(\zeta_{k+1}) > \frac{p-1}{2}$. Hence the result.

We can consider several generalized models belonging to the family of generalized type-1 Dirichlet family in the many matrix-variate cases. In all such situations we can derive the generalized Erdélyi-Kober fractional integral of the second kind in many matrices. We will take one such generalization here and obtain a theorem. Let $f_1(X_1, \ldots, X_k)$ be of the form

$$\begin{aligned} f_1(X_1, \ldots, X_k) &= C\ |X_1|^{\zeta_1} |I - X_1|^{\gamma_1} |X_2|^{\zeta_2} |I - X_1 - X_2|^{\gamma_2} \ldots \\ &\quad \times |X_k|^{\zeta_k} |I - X_1 - \ldots - X_k|^{\zeta_{k+1} + \gamma_k - \frac{p+1}{2}} \end{aligned} \tag{5.21}$$

where $O < X_1 + \ldots + X_j < I$, $j = 1, \ldots, k$, $\Re(\zeta_j) > -1$, $j = 1, \ldots, k$, $\Re(\zeta_{k+1}) > \frac{p-1}{2}$, C is the normalizing constant. Other conditions on the parameters will be given later. In this connection we can establish the following theorem.

Theorem 5.2 *Let $X_1, \ldots, X_k$ have a joint density as in* (5.21). *Consider the transformation as in* (5.15) *and* (5.16) *with the U_j's and V_j's defined as in* (5.17). *Let the joint density of $U_1, \ldots, U_k$ be again denoted as $g_2(U_1, \ldots, U_k)$. Let*

$$\delta_j = \zeta_{j+1} + \ldots + \zeta_{k+1} + \gamma_j + \gamma_{j+1} + \ldots + \gamma_k + (k - j)(\frac{p+1}{2}), \ j = 1, \ldots, k. \tag{5.22}$$

Then

$$\{\prod_{j=1}^{k} \frac{\Gamma_p(\zeta_j + \frac{p+1}{2})}{\Gamma_p(\zeta_j + \frac{p+1}{2} + \delta_j)}\} g_2(U_1, \ldots, U_k) = K_{2,U_j,\zeta_j,j=1,\ldots,k}^{-\delta_j,j=1,\ldots,k} f \tag{5.23}$$

where δ_j is defined in (5.22) *and the Erdélyi-Kober fractional integral of the second kind in* (5.14), $\Re(\zeta_j) > -1$, $\Re(\delta_j) > \frac{p-1}{2}$, $j = 1, \ldots, k$, $\Re(\zeta_{k+1}) > \frac{p-1}{2}$.

Proof We can show that under the transformation in (5.15) or (5.16) the Y_j's are independently distributed as real matrix-variate type-1 beta random variables with the parameters $(\zeta_j + \frac{p+1}{2}, \delta_j)$, $j = 1, \ldots, k$. Now the result follows from the procedure of the proof in Theorem 5.1.

Note 5.1 Special cases connecting to Riemann-Liouville fractional integral, Weyl fractional integral and Saigo fractional integral, corresponding to the ones in Sect. 5.1 for Erdélyi-Kober fractional integral of the second kind, are already mentioned in Chaps. 2 and 3, for one scalar variable or one matrix-variate case. These results can be extended to many scalar variables or many matrix variables case.

5.2 Fractional Integrals of the First Kind in the Case of Many Real Matrix Variables

Let $f_1(X_1, \ldots, X_k)$ be a function of many $p \times p$ real positive definite matrices and $f_2(V_1, \ldots, V_k)$ be another function of another sequence of $p \times p$ real positive definite matrices $V_1, \ldots, V_k$. Let f_1 be of the form

$$f_1(X_1, \ldots, X_k) = \{\prod_{j=1}^{k} \frac{\Gamma_p(\zeta_j + \alpha_j)}{\Gamma_p(\alpha_j)\Gamma_p(\zeta_j)} |X_j|^{\zeta_j - \frac{p+1}{2}} |I - X_j|^{\alpha_j - \frac{p+1}{2}}\} \tag{5.24}$$

for $O < X_j < I, \Re(\alpha_j) > \frac{p-1}{2}, \Re(\zeta_j) > \frac{p-1}{2}, j = 1, \ldots, k$. These are the real matrix-variate type-1 beta densities, with the parameters $(\zeta_j, \alpha_j), j = 1, \ldots, k$, on the right side. Let $U_j = V_j^{\frac{1}{2}} X_j^{-1} V_j^{\frac{1}{2}}, X_j = V^{\frac{1}{2}} U_j^{-1} V^{\frac{1}{2}}, j = 1, \ldots, k$. Let $f_2(V_1, \ldots, V_k) = f(V_1, \ldots, V_k)$ where f is arbitrary. Let the joint density of $U_1, \ldots, U_k$ be again denoted by $g_1(U_1, \ldots, U_k)$. Then

$$\begin{aligned}\{\prod_{j=1}^{k} \frac{\Gamma_p(\zeta_j)}{\Gamma_p(\zeta_j + \alpha_j)}\} g_1(U_1, \ldots, U_k) &= \{\prod_{j=1}^{k} \frac{1}{\Gamma_p(\alpha_j)} \int_{O<V_j<U_j} |V_j^{\frac{1}{2}} U_j^{-1} V_j^{\frac{1}{2}}|^{\zeta_j - \frac{p+1}{2}} \\ &\quad \times |I - V_j^{\frac{1}{2}} U_j^{-1} V_j^{\frac{1}{2}}|^{\alpha_j - \frac{p+1}{2}} |V_j|^{\frac{p+1}{2}} |U_j|^{-(p+1)}\} \\ &\quad \times f(V_1, \ldots, V_k) \mathrm{d}V_1 \wedge \ldots \wedge \mathrm{d}V_k \qquad (5.25) \\ &= K_{1,U_j,\zeta_j,j=1,\ldots,k}^{-\alpha_j,j=1,\ldots,k} f \qquad (5.26)\end{aligned}$$

for $\Re(\zeta_j) > \frac{p-1}{2}, \Re(\alpha_j) > \frac{p-1}{2}, j = 1, \ldots, k$. Then (5.26) will be taken as the definition of fractional integral of the first kind of orders $\alpha_1, \ldots, \alpha_k$, variables $U_1, \ldots, U_k$ and parameters $\zeta_1, \ldots, \zeta_k$ in the many matrix-variate case, where $g_1(U_1, \ldots, U_k)$ is a statistical density when $f(V_1, \ldots, V_k)$ is a statistical density.

Simplifying (5.25), we have a definition for generalized Erdélyi-Kober fractional integral of the first kind in many matrix variables case. We note that

$$K_{1,U_j,\zeta_j,j=1,\ldots,k}^{-\alpha_j,j=1,\ldots,k}f = \{\prod_{j=1}^{k}\frac{|U_j|^{-\zeta_j-\alpha_j}}{\Gamma_p(\alpha_j)}\int_{O<V_j<U_j}|V_j|^{\zeta_j}|U_j-V_j|^{\alpha_j-\frac{p+1}{2}}\}$$
$$\times f(V_1,\ldots,V_k)\mathrm{d}V_\wedge\ldots\wedge\mathrm{d}V_k, \quad (5.27)$$

for $\Re(\alpha_j) > \frac{p-1}{2}, \Re(\zeta_j) > \frac{p-1}{2}, j = 1,\ldots,k$. Hence from (5.24) to (5.27) we can have the following theorem.

Theorem 5.3 *Let $X_1,\ldots,X_k$ be independently distributed as real $p \times p$ matrix-variate type-1 beta random variables with parameters $(\zeta_j,\alpha_j), j = 1,\ldots,k, \Re(\alpha_j) > \frac{p-1}{2}, \Re(\zeta_j) > \frac{p-1}{2}$. Let $V_1,\ldots,V_k$ be another sequence of $p \times p$ real positive definite matrices having a joint density $f(V_1,\ldots,V_k)$. Let the two sets $(X_1,\ldots,X_k)$ and $(V_1,\ldots,V_k)$ be independently distributed. Let $U_j = V_j^{\frac{1}{2}}X_j^{-1}V_j^{\frac{1}{2}}, X_j = V_j^{\frac{1}{2}}U_j^{-1}V_j^{\frac{1}{2}}, j = 1,\ldots,k$. Let $g_1(U_1,\ldots,U_k)$ be the joint density of $U_1,\ldots,U_k$. Then the Erdélyi-Kober fractional integral of the first kind for many matrix variables case as defined in* (5.26) *is a constant multiple of $g_1(U_1,\ldots,U_k)$ as in* (5.25).

We can also have theorems parallel to the ones in Sect. 5.2 and the proofs are parallel. Hence we list two such theorems here without proofs.

Theorem 5.4 *Let $X_1,\ldots,X_k$ have a joint real $p \times p$ matrix-variate type-1 Dirichlet density with the parameters $(\zeta_1,\ldots,\zeta_k;\zeta_{k+1})$. Consider the transformation in* (5.15), (5.16) *and let $Y_1,\ldots,Y_k$ be as defined there. Let $V_1,\ldots,V_k$ be another sequence of $p \times p$ real positive definite matrix random variables having a joint density $f(V_1,\ldots,V_k)$ where let $(X_1,\ldots,X_k)$ and $(V_1,\ldots,V_k)$ be independently distributed. Let $U_j = V_j^{\frac{1}{2}}Y_j^{-1}V_j^{\frac{1}{2}}$, or $Y_j = V_j^{\frac{1}{2}}U_j^{-1}V_j^{\frac{1}{2}}, j = 1,\ldots,k$. Let the joint density of $U_1,\ldots,U_k$ be again denoted by $g_1(U_1,\ldots,U_k)$. Let*

$$\gamma_j = \zeta_{j+1}+\ldots+\zeta_{k+1}. \quad (5.28)$$

Then

$$g_1(U_1,\ldots,U_k) = \{\prod_{j=1}^{k}\frac{\Gamma_p(\zeta_j+\gamma_j)}{\Gamma_p(\zeta_j)}\}K_{1,U_j,\zeta_j,j=1,\ldots,k}^{-\gamma_j,j=1,\ldots,k}f \quad (5.29)$$

Or

$$\{\prod_{j=1}^{k}\frac{\Gamma_p(\zeta_j)}{\Gamma_p(\zeta_j+\gamma_j)}\}g_1(U_1,\ldots,U_k) = K_{1,U_j,\zeta_j,j=1,\ldots,k}^{-\gamma_j,j=1,\ldots,k}f, \quad (5.30)$$

for $\Re(\zeta_j) > \frac{p-1}{2}, j = 1,\ldots,k+1, \Re(\gamma_j) > \frac{p-1}{2}, j = 1,\ldots,k$.

Theorem 5.5 *Let $X_1, \ldots X_k$ have a joint density as in* (5.21) *with ζ_j replaced by $\zeta_j - \frac{p+1}{2}$ and the remaining transformations and notations remain as in Theorem* 5.3. *Let*

$$\delta_j = \zeta_{j+1} + \ldots + \zeta_{k+1} + \beta_j + \ldots + \beta_k. \tag{5.31}$$

Let the joint density of $U_1, \ldots, U_k$ be again denoted by $g_1(U_1, \ldots, U_k)$. Then $g_1(U_1, \ldots, U_k)$ is a density and

$$\{\prod_{j=1}^{k} \frac{\Gamma_p(\zeta_j)}{\Gamma_p(\zeta + \delta_j)}\} g_1(U_1, \ldots, U_k) = K_{1,U_j,\zeta_j,j=1,\ldots,k}^{-\delta_j,j=1,\ldots,k} f. \tag{5.32}$$

for $\Re(\zeta_j) > \frac{p-1}{2}, j = 1, \ldots, k+1, \Re(\delta_j) > \frac{p-1}{2}, j = 1, \ldots, k.$

5.3 M-Transforms for the Fractional Integrals in the Many Real Matrix-Variate Case

Here we look at the M-transforms for fractional integrals of the first and second kind in the many real matrix-variate case. Consider the first kind fractional integral in (5.27). The M-transform is given by

$$M\{K_{1,U_j,\zeta_j,j=1,\ldots,k}^{-\alpha_j,j=1,\ldots k} f; s_1, \ldots, s_k\} = \int_{U_1>O} \cdots \int_{U_k>O} |U_1|^{s_1 - \frac{p+1}{2}} \ldots |U_k|^{s_k - \frac{p+1}{2}}$$
$$\times K_{1,U_j,\zeta_j,j=1,\ldots,k}^{-\alpha_j,j=1,\ldots,k} f(U_1, \ldots, U_k) \mathrm{d}U_1 \wedge \ldots \wedge \mathrm{d}U_k$$

The U_j-integral is given by

$$\int_{U_j>V_j>O} |U_j|^{s_j - \frac{p+1}{2}} |U_j|^{-\zeta_j - \alpha_j} |U_j - V_j|^{\alpha_j - \frac{p+1}{2}} \mathrm{d}U_j$$
$$= |V_j|^{-\zeta_j + s_j - \frac{p+1}{2}} \int_{Y_j>O} |Y_j|^{\alpha_j - \frac{p+1}{2}} |I + Y_j|^{-(\zeta_j + \alpha_j - s_j + \frac{p+1}{2})} \mathrm{d}Y_j, \ T_j = U_j - V_j, Y_j = V_j^{-\frac{1}{2}} T_j V_j^{-\frac{1}{2}}$$
$$= |V_j|^{-\zeta_j + s_j - \frac{p+1}{2}} \frac{\Gamma_p(\alpha_j)\Gamma_p(\frac{p+1}{2} + \zeta_j - s_j)}{\Gamma_p(\frac{p+1}{2} + \zeta_j + \alpha_j - s_j)}$$

by evaluating the integral by using type-2 real matrix-variate beta integral, for $\Re(s) < \Re(\zeta_j + \frac{p+1}{2}), \Re(\alpha_j) > \frac{p-1}{2}, j = 1, \ldots, k$. Now the V_j-integrals give the M-transform of $f(V_1, \ldots, V_k)$. Hence we can have the following theorem.

Theorem 5.6 *For the Erdélyi-kober fractional integral of the first kind of orders* $\alpha_1, \ldots, \alpha_k$ *defined in* (5.30) *the M-transform is given by*

$$M\{K_{1,U_j,\zeta_j,j=1,\ldots,k}^{-\alpha_j,j=1,\ldots,k}f; s_1, \ldots, s_k\} = f^*(s_1, \ldots, s_k)\prod_{j=1}^{k}\frac{\Gamma_p(\frac{p+1}{2}+\zeta_j - s_j)}{\Gamma_p(\frac{p+1}{2}+\zeta_j+\alpha_j - s_j)}$$

for $\Re(s_j) < \Re(\zeta_j) + 1, \Re(\alpha_j) > \frac{p-1}{2}, j = 1, \ldots, k.$*, where* $f^*(s_1, \ldots, s_k)$ *is the M-transform of* $f(V_1, \ldots, V_k)$.

In a similar manner we can work out the M-transform of fractional integral of the second kind in the many real matrix-variate case. In this context we start with (5.13). The M-transform is given by

$$M\{K_{2,U_j,\zeta_j,j=1,\ldots,k}^{-\alpha_j,j=1,\ldots,k}f; s_1, \ldots, s_k\} = \{\frac{1}{\Gamma_p(\alpha_j)}\int_{U_j>O}|U_j|^{\zeta_j+s_j-\frac{p+1}{2}}\int_{V_j>U_j>O}|V_j|^{-\zeta_j-\alpha_j}$$

$$\times |V_j - U_j|^{\alpha_j-\frac{p+1}{2}}\mathrm{d}U_j\}f(V_1, \ldots, V_k)\mathrm{d}V_1 \wedge \ldots \wedge \mathrm{d}V_k$$

for $\Re(\alpha_j) > \frac{p-1}{2}, \Re(\zeta_j) > -1, j = 1, \ldots, k.$

The U_j-integral is given by

$$\int_{O<U_j<V_j}|U_j|^{s_j-\frac{p+1}{2}+\zeta_j}|V_j - U_j|^{\alpha_j-\frac{p+1}{2}}\mathrm{d}U_j$$

$$= |V_j|^{\alpha_j-\frac{p+1}{2}}\int_{O<U_j<V_j}|U_j|^{s_j+\zeta_j-\frac{p+1}{2}}|I - V^{-\frac{1}{2}}U_jV_j^{-\frac{1}{2}}|^{\alpha_j-\frac{p+1}{2}}\mathrm{d}U_j.$$

Put $Y_j = V^{-\frac{1}{2}}U_jV_j^{-\frac{1}{2}}$ and integrate out by using a real matrix-variate type-1 beta integral then the U_j integral is

$$= |V_j|^{\alpha_j-\frac{p+1}{2}+s_j+\zeta_j}\frac{\Gamma_p(\alpha_j)\Gamma_p(\zeta_j+s_j)}{\Gamma_p(\alpha_j+\zeta_j+s_j)}$$

for $\Re(\alpha_j) > \frac{p-1}{2}, \Re(\zeta_j + s_j) > \frac{p-1}{2}$. Now the V_j-integrals give the M-transform of f. Hence we have the following theorem.

Theorem 5.7 *For the Erdélyi-Kober fractional integral of the second kind defined in* (5.13) *the M-transform is given by*

$$M\{K_{2,U_j,\zeta_j,j=1,\ldots,k}^{-\alpha_j,j=1,\ldots,k}f; s_1, \ldots, s_k\} = f^*(s_1, \ldots, s_k)\prod_{j=1}^{k}\frac{\Gamma_p(\zeta_j+s_j)}{\Gamma_p(\alpha_j+\zeta_j+s_j)}$$

for $\Re(\alpha_j) > \frac{p-1}{2}, \Re(\zeta_j + s_j) > \frac{p-1}{2}, j = 1, \ldots, k$, *where* $f^*(s_1, \ldots, s_k)$ *is the M-transform of* $f(V_1, \ldots, V_k)$.

Note that for $k = 1$ the corresponding M-transforms in the one real matrix variable case, for the first and second kind fractional integrals are available from Theorems 4.7 and 4.5. For $p = 1$ the corresponding Mellin transforms in the k real scalar variables case and for $p = 1, k = 1$ the corresponding Mellin transforms in the one real scalar variable case for the Erdélyi-Kober fractional integral of the first and second kinds are obtained from Theorems 3.5 and 3.3 respectively.

References

1. K.M. Kurian, B. Kurian, A.M. Mathai, A matrix-variate extension of inverted Dirichlet integral. Proc. Natl. Acad. Sci. (India) **74(A)II**, 1–10 (2014)
2. A.M. Mathai, Fractional integral operators involving many matrix variables. Linear Algebra Appl. **446**, 196–215 (2014)
3. A.M. Mathai, S.B. Provost, Various generalizations to the Dirichlet distribution. Stat. Methods **8(2)**, 142–163 (2006)

Chapter 6
Erdélyi-Kober Fractional Integrals in the Complex Domain

6.1 Introduction

In Chaps. 2, 3, 4, and 5 we considered the real scalar variable case, real multivariate case, real one matrix-variate case, real several matrix-variate case. In the present chapter we will look into fractional calculus in the complex domain. Since we will be dealing with $p \times p$ Hermitian positive definite matrices, for $p = 1$ Hermitian positive definite means a real scalar positive variable. Hence we start with $p \geq 2$. Fractional calculus of one real scalar variable case is the one most frequently appearing in various theoretical and applied areas. Fractional calculus in the complex domain was considered only recently, see Mathai [2]. The following discussion is based on this work.

The following are the standard notations which will be used in the present and succeeding chapters. All matrices appearing are $p \times p$ with elements in the complex domain unless stated otherwise. $\det(\cdot)$ will denote the determinant of $(\cdot)$. $|\det(\cdot)|$ will be the absolute value of the determinant of $(\cdot)$. A matrix X with scalar complex variables as elements will be denoted by a tilde as $\tilde{X}$. Constant matrices will not be written with a tilde whether in the real or complex domain. $\mathrm{tr}(X)$ is the trace of X, $(\mathrm{d}\tilde{X}) = (\mathrm{d}\tilde{x}_{ij})$ is the matrix of differentials $\mathrm{d}\tilde{x}_{ij}$'s. Let $\tilde{X} = X_1 + iX_2$ where X_1 and X_2 are real $m \times n$ matrices and $i = \sqrt{(-1)}$. Then $\mathrm{d}\tilde{X} = \mathrm{d}X_1 \wedge \mathrm{d}X_2$ where

$$\mathrm{d}X_1 = \prod_{i=1}^{m}\prod_{j=1}^{n} \wedge \mathrm{d}x_{ij1} \text{ and } \mathrm{d}X_2 = \prod_{i=1}^{m}\prod_{j=1}^{n} \wedge \mathrm{d}x_{ij2}$$

where x_{ij1} and x_{ij2} are the $(i, j) - th$ elements in X_1 and X_2 respectively, and $\wedge$ denotes the wedge product. For any $p \times p$ matrix $A = A_1 + iA_2$ in the complex domain, the determinant will be a complex number of the form $\det(A) = a + ib$ where a and b are real scalar quantities. Then the absolute value of the determinant will be of the form $|\det(A)| = +[(a + ib)(a - ib)]^{\frac{1}{2}} = +[a^2 + b^2]^{\frac{1}{2}}$. Note that the

A. M. Mathai, H. J. Haubold, *Erdélyi–Kober Fractional Calculus*, SpringerBriefs in Mathematical Physics 31, https://doi.org/10.1007/978-981-13-1159-8_6

conjugate of $A_1 + iA_2$ is $A_1 - iA_2$. Then the square of the absolute value of the determinant of A can also be written as the following determinant, where $\bar{A}$ denotes the conjugate of A:

$$\begin{aligned}|\det(A)|^2 &= \det(A)\det(\bar{A}) = \det(A_1 + iA_2)\det(A_1 - iA_2)\\ &= \det\begin{bmatrix} A_1 + iA_2 & O \\ O & A_1 - iA_2 \end{bmatrix}.\end{aligned} \tag{6.1}$$

The block diagonal determinant on the right can be brought to the following forms:

$$\begin{aligned}\det\begin{bmatrix} A_1 + iA_2 & O \\ O & A_1 - iA_2 \end{bmatrix} &= \det(B) = \det(C),\\ B = \begin{bmatrix} A_1 & A_2 \\ -A_2 & A_1 \end{bmatrix}&, \; C = \begin{bmatrix} A_1 & -A_2 \\ A_2 & A_1 \end{bmatrix}.\end{aligned} \tag{6.2}$$

Hence we have

$$|\det(A)| = |\det(B)|^{\frac{1}{2}} = |\det(C)|^{\frac{1}{2}}. \tag{6.3}$$

We need a few basic results on Jacobians of matrix transformations in the complex domain. These and further properties may be seen from Mathai [1]. The results that we need will be listed as lemmas here without proofs.

Lemma 6.1 *Let $\tilde{X}$ and $\tilde{Y}$ be $m \times n$ matrices in the complex domain. Let A be $m \times m$ and B be $n \times n$ nonsingular constant matrices in the sense of free of the elements in $\tilde{X}$ and $\tilde{Y}$. Let C be a constant $m \times n$ matrix. Then*

$$\tilde{Y} = A\tilde{X}B + C, \det(A) \neq 0, \det(B) \neq 0 \Rightarrow \mathrm{d}\tilde{Y} = |\det(AA^*)|^n |\det(BB^*)|^m \mathrm{d}\tilde{X}, \tag{6.4}$$

where A^* and B^* denote the conjugate transposes of A and B respectively.

When $A = A^*$, that is, when a matrix A in the complex domain is equal to its conjugate transpose then it is called a Hermitian matrix. The next result is about the transformation of a Hermitian matrix to a Hermitian matrix.

Lemma 6.2 *Let $\tilde{X}$ and $\tilde{Y}$ be $p \times p$ Hermitian matrices and let A be a nonsingular constant matrix. Then*

$$\tilde{Y} = A\tilde{X}A^* \Rightarrow \mathrm{d}\tilde{Y} = \begin{cases} |\det(A)|^{2p}\mathrm{d}\tilde{X} \\ |\det(AA^*)|^{p}\mathrm{d}\tilde{X} \end{cases} \tag{6.5}$$

The next result is on a decomposition of the Hermitian positive definite matrix $\tilde{X} = \tilde{X}^* > O$.

Lemma 6.3 *Let $\tilde{X}$ be a $p \times p$ Hermitian positive definite matrix. Let $\tilde{T}$ be a $p \times p$ lower triangular matrix with diagonal elements t_{jj}'s being real and positive. Consider the unique representation $\tilde{X} = \tilde{T}\tilde{T}^*$. Then*

$$\tilde{X} = \tilde{T}\tilde{T}^* \Rightarrow \mathrm{d}\tilde{X} = 2^p\{\prod_{j=1}^{p} t_{jj}^{2(p-j)+1}\}\mathrm{d}\tilde{T}. \tag{6.6}$$

Next we define a complex matrix-variate gamma function, denoted by $\tilde{\Gamma}_p(\alpha)$, as

$$\tilde{\Gamma}_p(\alpha) = \pi^{\frac{p(p-1)}{2}}\Gamma(\alpha)\Gamma(\alpha-1)\ldots\Gamma(\alpha-p+1),\ \Re(\alpha) > p-1. \tag{6.7}$$

This complex matrix-variate gamma has the following integral representation:

$$\tilde{\Gamma}_p(\alpha) = \int_{\tilde{X}>O} |\det(\tilde{X})|^{\alpha-p}\mathrm{e}^{-\mathrm{tr}(\tilde{X})}\mathrm{d}\tilde{X}, \Re(\alpha) > p-1. \tag{6.8}$$

This can be established by using Lemma 6.3. With the help of (6.8) and Lemma 6.2 we can define a matrix-variate gamma density in the complex domain as follows:

$$f(\tilde{X}) = \frac{|\det(B)|^\alpha}{\tilde{\Gamma}_p(\alpha)}|\det(\tilde{X})|^{\alpha-p}\mathrm{e}^{-\mathrm{tr}(B\tilde{X})},\ \tilde{X} = \tilde{X}^* > O,\ \Re(\alpha) > p-1 \tag{6.9}$$

and $f(\tilde{X}) = 0$ elsewhere, where $B = B^* > O$ is a constant Hermitian positive definite matrix.

Lemma 6.4 *Let $\tilde{X}$ be a nonsingular matrix and let $\tilde{Y} = \tilde{X}^{-1}$. Then*

$$\tilde{Y} = \tilde{X}^{-1} \Rightarrow \mathrm{d}\tilde{Y} = \begin{cases} |\det(\tilde{X}\tilde{X}^*)|^{-2p}\mathrm{d}\tilde{X} \text{ for a general } \tilde{X} \\ |\det(\tilde{X}\tilde{X}^*)|^{-p} \text{ for } \tilde{X} = \tilde{X}^* \text{ or } \tilde{X} = -\tilde{X}^*. \end{cases} \tag{6.10}$$

Lemma 6.5 *Let $\tilde{Y}$ be $p \times n, n \geq p$ and of full rank p. Let $\tilde{Y}\tilde{Y}^* = \tilde{S}$. Then after integrating over the Stiefel manifold*

$$\mathrm{d}\tilde{Y} = \frac{\pi^{np}}{\tilde{\Gamma}_p(n)}|\det(\tilde{S})|^{n-p}\mathrm{d}\tilde{S},$$

see Mathai ([1], Corollaries 4.5.2, 4.5.3).

We need complex matrix-variate beta function and its integral representations. The complex matrix-variate beta function will be denoted and defined as follows:

$$\tilde{B}_p(\alpha,\beta) = \frac{\tilde{\Gamma}_p(\alpha)\tilde{\Gamma}_p(\beta)}{\tilde{\Gamma}_p(\alpha+\beta)}, \Re(\alpha) > p-1, \Re(\beta) > p-1 \tag{6.11}$$

$$= \int_{O<\tilde{X}<I} |\det(\tilde{X})|^{\alpha-p}|\det(I-\tilde{X})|^{\beta-p}\mathrm{d}\tilde{X}, \text{ (type-1)} \tag{6.12}$$

$$= \int_{\tilde{U}>0} |\det(\tilde{U})|^{\alpha-p}|\det(I+\tilde{U})|^{-(\alpha+\beta)}\mathrm{d}\tilde{U}, \text{ (type-2)} \tag{6.13}$$

for $\Re(\alpha) > p-1, \Re(\beta) > p-1$ where, in general, $\int_{A<\tilde{X}<B} f(\tilde{X})\mathrm{d}\tilde{X}$ will mean the integral of a real-valued scalar function $f(\tilde{X})$ of complex matrix argument $\tilde{X}$ and the integral is taken over all $\tilde{X}$ such that $A = A^* > O, B = B^* > O, \tilde{X} = \tilde{X}^* > O, \tilde{X} - A > O, B - \tilde{X} > O$, where A and B are constant matrices.

6.2 Explicit Evaluations of Gamma and Beta Integrals in the Complex Domain

We have seen from (6.8) that

$$\tilde{\Gamma}_p(\alpha) = \int_{\tilde{X}>O} |\det(\tilde{X})|^{\alpha-p}\mathrm{e}^{-\mathrm{tr}(\tilde{X})}\mathrm{d}\tilde{X}. \tag{6.14}$$

One standard procedure to evaluate the integral in (6.14) is to write the Hermitian positive definite matrix as $\tilde{X} = \tilde{T}\tilde{T}^*$ where $\tilde{T}$ is a lower triangular matrix with real and positive diagonal elements $t_{jj} > 0, j = 1, \ldots, p$, where $*$ indicates the conjugate transpose. Then the Jacobian is available from Lemma 6.3. Then

$$\begin{aligned}\mathrm{tr}(\tilde{X}) &= \mathrm{tr}(\tilde{T}\tilde{T}^*) \\ &= t_{11}^2 + \ldots + t_{pp}^2 + |\tilde{t_{21}}|^2 + \ldots + |\tilde{t_{p1}}|^2 + \ldots + |\tilde{t_{pp-1}}|^2\end{aligned}$$

and

$$\mathrm{d}\tilde{X} = 2^p\{\prod_{j=1}^{p} t_{jj}^{2\alpha-2j+1}\}\mathrm{d}\tilde{T}.$$

Now, integrating out over $\tilde{t_{jk}}$ for $j > k$

$$\int_{\tilde{t_{jk}}} \mathrm{e}^{-|\tilde{t_{jk}}|^2}\mathrm{d}\tilde{t_{jk}} = \int_{-\infty}^{\infty}\int_{-\infty}^{\infty} \mathrm{e}^{-(t_{jk1}^2+t_{jk2}^2)}\mathrm{d}t_{jk1}\wedge\mathrm{d}t_{jk2} = \pi$$

and

$$\prod_{j>k}\pi = \pi^{\frac{p(p-1)}{2}}.$$

Now,

$$2\int_0^\infty t_{jj}^{2\alpha-2j+1}\mathrm{e}^{-t_{jj}^2}\mathrm{d}t_{jj} = \Gamma(\alpha-j+1),\ \Re(\alpha) > j-1,$$

for $j = 1, \ldots, p$. Now the product of all these gives

$$\pi^{\frac{p(p-1)}{2}}\Gamma(\alpha)\Gamma(\alpha-1)\ldots\Gamma(\alpha-p+1) = \tilde{\Gamma}_p(\alpha),\ \Re(\alpha) > p-1$$

and hence the result is verified.

6.2.1 An Alternate Method Based on Partitioned Matrix

Let us separate x_{pp}. When $\tilde{X}$ is $p\times p$ Hermitian positive definite then all its diagonal elements are real and positive. That is, $x_{jj} > 0,\ j = 1, \ldots, p$. Let

$$\tilde{X} = \begin{bmatrix}\tilde{X_{11}} & \tilde{X_{12}}\\ \tilde{X_{21}} & x_{pp}\end{bmatrix}$$

where $\tilde{X_{11}}$ is $(p-1)\times(p-1)$ and

$$|\det(\tilde{X})|^{\alpha-p} = |\det(\tilde{X_{11}})|^{\alpha-p}|x_{pp} - \tilde{X_{21}}\tilde{X_{11}}^{-1}\tilde{X_{12}}|^{\alpha-p}$$

and

$$\mathrm{tr}(\tilde{X}) = \mathrm{tr}(\tilde{X_{11}}) + x_{pp}.$$

Then

$$|x_{pp} - \tilde{X_{21}}\tilde{X_{11}}^{-1}\tilde{X_{12}}|^{\alpha-p} = x_{pp}^{\alpha-p}|1 - x_{pp}^{-\frac{1}{2}}\tilde{X_{21}}\tilde{X_{11}}^{-\frac{1}{2}}\tilde{X_{11}}^{-\frac{1}{2}}\tilde{X_{12}}x_{pp}^{-\frac{1}{2}}|^{\alpha-p}.$$

Put

$$\tilde{Y} = x_{pp}^{-\frac{1}{2}}\tilde{X_{21}}\tilde{X_{11}}^{-\frac{1}{2}} \Rightarrow \mathrm{d}\tilde{Y} = x_{pp}^{-(p-1)}|\det(\tilde{X_{11}})|^{-1}\mathrm{d}\tilde{X_{21}}$$

from Lemma 6.1. Now, the integral over x_{pp} gives

$$\int_0^\infty x_{pp}^{\alpha-p+(p-1)}\mathrm{e}^{-x_{pp}}\mathrm{d}x_{pp} = \Gamma(\alpha),\ \Re(\alpha) > 0.$$

Let $u = \tilde{Y}\tilde{Y}^*$. Then $\mathrm{d}\tilde{Y} = u^{p-2}\frac{\pi^{p-1}}{\Gamma(p-1)}\mathrm{d}u$ from Lemma 6.5

$$\int_0^\infty u^{(p-1)-1}(1-u)^{\alpha-(p-1)-1}\mathrm{d}u = \frac{\Gamma(p-1)\Gamma(\alpha-(p-1))}{\Gamma(\alpha)}, \Re(\alpha) > p-1.$$

Taking the product we have

$$|\det(\tilde{X}_{11}^{(1)})|^{\alpha+1-p}\Gamma(\alpha)\frac{\pi^{p-1}}{\Gamma(p-1)}\frac{\Gamma(p-1)\Gamma(\alpha-(p-1))}{\Gamma(\alpha)}$$
$$= \pi^{p-1}\Gamma(\alpha-(p-1))|\det(\tilde{X}_{11}^{(1)})|^{\alpha+1-p}$$

where $\tilde{X}_{11}^{(1)}$ indicates $\tilde{X}_{11}$ after the first set of integrations. Now for the second stage, separate $x_{p-1,p-1}$ and the first $(p-2)\times(p-2)$ block may be denoted by $\tilde{X}_{11}^{(2)}$. Now proceed as before to get $|\det(\tilde{X}_{11}^{(2)})|^{\alpha+2-p}\pi^{p-2}\Gamma(\alpha-(p-2))$. Proceeding like this we have the exponent of π as $(p-1)+(p-2)+\ldots+1 = p(p-1)/2$ and the gamma product will be $\Gamma(\alpha-(p-1))\Gamma(\alpha-(p-2))\ldots\Gamma(\alpha)$ for $\Re(\alpha) > p-1$. That is,

$$\pi^{\frac{p(p-1)}{2}}\Gamma(\alpha)\Gamma(\alpha-1)\ldots\Gamma(\alpha-(p-1)) = \tilde{\Gamma}_p(\alpha).$$

6.3 Evaluation of Matrix-Variate Beta Integrals in the Complex Domain

Here we will consider a direct way of evaluating matrix-variate type-1 and type-2 beta integrals in the real and complex cases, see also Mathai [3].

One integral representation for $\tilde{B}_p(\alpha, \beta)$ in the complex case is the following:

$$\int_{O<\tilde{X}<I}|\det(\tilde{X})|^{\alpha-p}|\det(I-\tilde{X})|^{\beta-p}\mathrm{d}\tilde{X} = \tilde{B}_p(\alpha, \beta)$$

for $\Re(\alpha) > p-1,\ \Re(\beta) > p-1$ where $\det(\cdot)$ denotes the determinant of $(\cdot)$ and $|\det(\cdot)|$ denotes the absolute value of the determinant of $(\cdot)$. Here $\tilde{X} = (\tilde{x}_{ij})$ is a $p\times p$ Hermitian positive definite matrix and hence all the diagonal elements are real and positive. As in the real case, let us separate x_{pp} by partitioning:

$$\tilde{X} = \begin{bmatrix}\tilde{X}_{11} & \tilde{X}_{12}\\ \tilde{X}_{21} & \tilde{X}_{22}\end{bmatrix} \text{ as well as } I-\tilde{X} = \begin{bmatrix}I-\tilde{X}_{11} & -\tilde{X}_{12}\\ -\tilde{X}_{21} & I-\tilde{X}_{22}\end{bmatrix}.$$

Then the absolute value of the determinants are of the form:

$$|\det(\tilde{X})|^{\alpha-p} = |\det(\tilde{X}_{11})|^{\alpha-p}|x_{pp} - \tilde{X}_{21}\tilde{X}_{11}^{-1}\tilde{X}_{12}^{*}|^{\alpha-p} \quad (i)$$

where * indicates conjugate transpose, and

$$|\det(I-\tilde{X})|^{\beta-p} = |\det(I-\tilde{X}_{11})|^{\beta-p}|(1-x_{pp})-\tilde{X}_{21}(I-\tilde{X}_{11})^{-1}\tilde{X}_{12}^{*}|^{\beta-p}. \quad (ii)$$

Note that when $\tilde{X}$ and $I - \tilde{X}$ are Hermitian positive definite then $\tilde{X}_{11}^{-1}$ and $(I - \tilde{X}_{11})^{-1}$ are also Hermitian positive definite. Further, the Hermitian forms $\tilde{X}_{21}\tilde{X}_{11}^{-1}\tilde{X}_{12}^{*}$ and $\tilde{X}_{21}(I - \tilde{X}_{11})^{-1}\tilde{X}_{12}^{*}$ remain real and positive. From (i) and (ii) it follows that

$$\tilde{X}_{21}\tilde{X}_{11}^{-1}\tilde{X}_{12}^{*} < x_{pp} < 1 - \tilde{X}_{21}(I - \tilde{X}_{11})^{-1}\tilde{X}_{12}^{*}.$$

Since Hermitian forms are real, the lower and upper bounds of x_{pp} are real. Let

$$\tilde{W} = \tilde{X}_{21}\tilde{X}_{11}^{-\frac{1}{2}}(I - \tilde{X}_{11})^{-\frac{1}{2}}$$

for fixed $\tilde{X}_{11}$. Then

$$\mathrm{d}\tilde{X}_{21} = |\det(\tilde{X}_{11})|^{-1}|\det(I - \tilde{X}_{11})|^{-1}\mathrm{d}\tilde{W}$$

and $|\det(\tilde{X})|^{\alpha-p}, |\det(I - \tilde{X}_{11})|^{\beta-p}$ change to $|\det(\tilde{X}_{11})|^{\alpha+1-p}, |\det(I - \tilde{X}_{11})|^{\beta+1-p}$ respectively. Then we can write: $y = x_{pp} - \tilde{X}_{21}\tilde{X}_{11}^{-1}\tilde{X}_{12}$, $b = 1 - \tilde{X}_{21}\tilde{X}_{11}^{-1}\tilde{X}_{12} - \tilde{X}_{21}(I - \tilde{X}_{11})^{-1}\tilde{X}_{12}$ so that

$$|(1 - x_{pp}) - \tilde{X}_{21}\tilde{X}_{11}^{-1}\tilde{X}_{12}^{*} - \tilde{X}_{21}(I - \tilde{X}_{11})^{-1}\tilde{X}_{12}^{*}|^{\beta-p} = (b - y)^{\beta-p} = b^{\beta-p}[1 - \frac{y}{b}]^{\beta-p}.$$

Put $u = \frac{y}{b}$. Then the factors containing u and b will be of the form $u^{\alpha-p}(1 - u)^{\beta-p}b^{\alpha+\beta-2p+1}$ and the integral over u gives

$$\int_0^1 u^{\alpha-p}(1 - u)^{\beta-p}\mathrm{d}u = \frac{\Gamma(\alpha - (p - 1))\Gamma(\beta - (p - 1))}{\Gamma(\alpha + \beta - 2(p - 1))},$$

for $\Re(\alpha) > p - 1, \Re(\beta) > p - 1$. Let $v = \tilde{W}\tilde{W}^{*}$. Then from Lemma 6.5

$$\mathrm{d}\tilde{W} = \frac{\pi^{p-1}}{\Gamma(p - 1)}v^{(p-1)-1}\mathrm{d}v.$$

The integral over b gives

$$\int b^{\alpha+\beta-2p+1}\mathrm{d}\tilde{X}_{21} = \int_0^1 v^{(p-1)-1}(1-v)^{\alpha+\beta-2p+1}\mathrm{d}v = \frac{\Gamma(p-1)\Gamma(\alpha+\beta-2p+2)}{\Gamma(\alpha+\beta-p+1)},$$

for $\Re(\alpha) > p-1, \Re(\beta) > p-1$. Now, taking the product of all factors we have

$$|\det(\tilde{X}_{11})|^{\alpha+1-p}|\det(I-\tilde{X}_{11})|^{\beta+1-p}\pi^{p-1}\frac{\Gamma(\alpha-p+1)\Gamma(\beta-p+1)}{\Gamma(\alpha+\beta-p+1)}$$

for $\Re(\alpha) > p-1, \Re(\beta) > p-1$. Separate $x_{p-1,p-1}$ from $\tilde{X}_{11}$ and $I-\tilde{X}_{11}$ and continue the process. Then at the end, the exponent of π will be $(p-1)+(p-2)+\ldots+1 = \frac{p(p-1)}{2}$ and the gamma product will be

$$\frac{\Gamma(\alpha-(p-1))\Gamma(\alpha-(p-2))\ldots\Gamma(\alpha)\Gamma(\beta-(p-1))\ldots\Gamma(\beta)}{\Gamma(\alpha+\beta-(p-1))\ldots\Gamma(\alpha+\beta)}.$$

These factors, together with $\pi^{\frac{p(p-1)}{2}}$ give

$$\frac{\tilde{\Gamma}_p(\alpha)\tilde{\Gamma}_p(\beta)}{\tilde{\Gamma}_p(\alpha+\beta)} = \tilde{B}_p(\alpha,\beta), \Re(\alpha) > p-1, \Re(\beta) > p-1.$$

The procedure for evaluating a type-2 matrix-variate beta integral by the method of partitioning is parallel and hence it will not be detailed here. We can also consider a general partitioning as in the real case. Also, we can consider a method of avoiding the integration over the Stiefel manifold and the steps are parallel to those in the real case and hence deleted.

6.4 Fractional Integrals in the Matrix-Variate Case in the Complex Domain

We will introduce a general definition of what is meant by fractional integrals in the complex matrix-variate case. This definition will be introduced in terms of M-convolutions of products and ratios or convolutions in terms of generalized matrix transforms or M-transforms discussed in Mathai [1]. It is easy to introduce the concepts in terms of statistical densities of products and ratios of matrices in the complex domain. This will also give a physical interpretation of M-transforms. In order to illustrate the concepts let us look at the problem of deriving the density of a product of two matrix-variate random variables in the complex domain. Let $\tilde{U}_2 = \tilde{X}_2^{\frac{1}{2}}\tilde{X}_1\tilde{X}_2^{\frac{1}{2}}$ where $\tilde{U}_2, \tilde{X}_1, \tilde{X}_2$ are $p\times p$ matrices in the complex domain. This $\tilde{U}_2$ will be interpreted as the symmetric product of matrices. Here $\tilde{X}_2^{\frac{1}{2}}$ denotes the Hermitian positive definite square root of $\tilde{X}_2$. When $\tilde{X}_2$ is Hermitian positive definite all its eigenvalues are real and positive and there exists a unitary matrix $\tilde{Z}$, $\tilde{Z}^*\tilde{Z} = I, \tilde{Z}\tilde{Z}^* = I$ such that $\tilde{X}_2 = \tilde{Z}^* D\tilde{Z}, D = \text{diag}(\lambda_1, \ldots, \lambda_p)$ where $\lambda_j > 0, j = 1, \ldots, p$ are the eigenvalues of $\tilde{X}_2$. Then $\tilde{X}_2^{\frac{1}{2}} = \tilde{Z}^* D^{\frac{1}{2}}\tilde{Z}, D^{\frac{1}{2}} = \text{diag}(\lambda_1^{\frac{1}{2}}, \ldots, \lambda_p^{\frac{1}{2}})$. Let $\tilde{X}_1 = \tilde{X}_1^* > O$ and $\tilde{X}_2 = \tilde{X}_2^* > O$ be Hermitian positive definite with the densities $f_1(\tilde{X}_1)$ and $f_2(\tilde{X}_2)$ respectively. Here, a density means a real-valued scalar function of matrix argument with the matrix in the complex domain, $f(\tilde{X})$, such that $f(\tilde{X}) \geq 0$ for all $\tilde{X}$ and the total integral $\int_{\tilde{X}} f(\tilde{X})\text{d}\tilde{X} = 1$. Let $\tilde{X}_1$ and $\tilde{X}_2$ be independently distributed and let the density of $\tilde{U}_2$ be denoted by $g_2(\tilde{U}_2)$. Then the joint density of $\tilde{X}_1$ and $\tilde{X}_2$, denoted by $f(\tilde{X}_1, \tilde{X}_2)$, is the product of marginal densities due to statistical independence. That is, $f(\tilde{X}_1, \tilde{X}_2) = f_1(\tilde{X}_1) f_2(\tilde{X}_2)$. Consider the transformation $\tilde{U}_2 = \tilde{X}_2^{\frac{1}{2}}\tilde{X}_1\tilde{X}_2^{\frac{1}{2}}, \tilde{X}_2 = \tilde{V}$ so that $\tilde{X}_2 = \tilde{V}$ and $\tilde{X}_1 = \tilde{V}^{-\frac{1}{2}}\tilde{U}_2\tilde{V}^{-\frac{1}{2}}$. Then from Lemma 6.2 the Jacobian of this transformation is given by

$$\text{d}\tilde{X}_1 \wedge \text{d}\tilde{X}_2 = |\text{det}(\tilde{V})|^{-p}\text{d}\tilde{U}_2 \wedge \text{d}\tilde{V} \tag{6.15}$$

and then the density of $\tilde{U}_2$ is given by

$$\tilde{g}_2(\tilde{U}_2) = \int_{\tilde{V}} |\text{det}(\tilde{V})|^{-p} f_1(\tilde{V}^{-\frac{1}{2}}\tilde{U}_2\tilde{V}^{-\frac{1}{2}}) f_2(\tilde{V})\text{d}\tilde{V}. \tag{6.16}$$

Note that when f_1 and f_2 are statistical densities then $\tilde{g}_2(\tilde{U}_2)$ is the statistical density of the product $\tilde{U}_2 = \tilde{X}_2^{\frac{1}{2}}\tilde{X}_1\tilde{X}_2^{\frac{1}{2}}$. If f_1 and f_2 are arbitrary functions, need not be densities, then $\tilde{g}_2(\tilde{U}_2)$ of (6.16) will be called the M-convolution of a product and when $p = 1$, (6.16) is nothing but the Mellin convolution of a product. It is trivial to note that the M-transform of $\tilde{g}_2$ in (6.16) is the product of the M-transforms of f_1 and f_2. This will be stated as Theorem 6.1. The M-transform of a real-valued scalar function $f(\tilde{X})$ with parameter s, where $\tilde{X} = \tilde{X}^* > O$ is a $p \times p$ Hermitian positive definite matrix in the complex domain, is defined as follows, when the integral is convergent (see Mathai [1]):

$$\tilde{M}_f(s) = \tilde{M}\{f; s\} = \int_{\tilde{X}>O} |\text{det}(\tilde{X})|^{s-p} f(\tilde{X})\text{d}\tilde{X}. \tag{6.17}$$

Theorem 6.1 *For the M-transform defined in* (6.17)*, the M-transform of* $\tilde{g}_2$ *of* (6.16) *is given by*

$$\tilde{M}_{g_2}(s) = \tilde{M}_{f_1}(s)\tilde{M}_{f_2}(s). \tag{6.18}$$

Proof Taking the M-transform on both sides, with parameter s, we have

$$\tilde{M}_{g_2}(s) = \int_{\tilde{U}_2>O} |\det(\tilde{U}_2)|^{s-p}[\int_{\tilde{V}} |\det(\tilde{V})|^{-p} f_1(\tilde{V}^{-\frac{1}{2}}\tilde{U}_2\tilde{V}^{-\frac{1}{2}}) f_2(\tilde{V})\mathrm{d}\tilde{V}]\mathrm{d}\tilde{U}_2$$

$$= \int_{\tilde{V}} f_2(\tilde{V})[\int_{\tilde{U}} |\det(\tilde{U}_2)|^{s-p}|\det(\tilde{V})|^{-p} f_1(\tilde{V}^{-\frac{1}{2}}\tilde{U}_2\tilde{V}^{-\frac{1}{2}})\mathrm{d}\tilde{U}_2]\mathrm{d}\tilde{V} \qquad \text{(a)}$$

Now, put $\tilde{W} = \tilde{V}^{-\frac{1}{2}}\tilde{U}_2\tilde{V}^{-\frac{1}{2}} \Rightarrow \mathrm{d}\tilde{U}_2 = |\det(\tilde{V})|^p\mathrm{d}\tilde{W}$. Then the right side in (a) becomes

$$\int_{\tilde{W}} |\det(\tilde{W})|^{s-p} f_1(\tilde{W})\mathrm{d}\tilde{W} \int_{\tilde{V}} |\det(\tilde{V})|^{s-p} f_2(\tilde{V})\mathrm{d}\tilde{V} = \tilde{M}_{f_1}(s)\tilde{M}_{f_2}(s)$$

and hence the result.

Now, we can give a formal definition of fractional integrals in the complex matrix-variate case.

Definition 6.1 (Fractional Integral of the Second Kind of Order α in the Complex Matrix-variate Case) Fractional integral of the second kind of order α in the complex matrix-variate case will be defined as the M-convolution of a product as in (6.16) where

$$f_1(\tilde{X}_1) = \phi_1(\tilde{X}_1)\frac{|\det(I - \tilde{X}_1)|^{\alpha-p}}{\tilde{\Gamma}_p(\alpha)}$$

for $\Re(\alpha) > p - 1$, where $\phi_1(\tilde{X}_1)$ is a specified function, and

$$f_2(\tilde{X}_2) = \phi_2(\tilde{X}_2)f(\tilde{X}_2)$$

where $\phi_2(\tilde{X}_2)$ is a specified function and $f(\tilde{X}_2)$ is an arbitrary function.

Note 6.1 In the real matrix-variate case this definition will become as follows: $f_2(X_2) = \phi_2(X_2)f(X_2)$ where ϕ_2 is a specified function and $f(X_2)$ is an arbitrary function, and

$$f_1(X_1) = \phi_1(X_1)\frac{1}{\Gamma_p(\alpha)}[\det(I - X_1)]^{\alpha-\frac{p+1}{2}}, \ \Re(\alpha) > \frac{p-1}{2}$$

where $\phi_1(X_1)$ is a specified function, $\Gamma_p(\alpha)$ is the real matrix-variate gamma function defined as

$$\Gamma_p(\alpha) = \pi^{\frac{p(p-1)}{4}}\Gamma(\alpha)\Gamma(\alpha - \frac{1}{2})\ldots\Gamma(\alpha - \frac{p-1}{2})\Re(\alpha) > \frac{p-1}{2}. \qquad (6.19)$$

6.4.1 Erdélyi-Kober Fractional Integral of the Second Kind of Order α

Let $\phi_2(\tilde{X}_2) = 1$ and

$$\phi_1(\tilde{X}_1) = \frac{\tilde{\Gamma}_p(\alpha+\beta+p)}{\tilde{\Gamma}_p(\beta+p)}|\det(\tilde{X}_1)|^{\beta}.$$

In this case, $f_1(\tilde{X}_1)$ is a complex matrix-variate type-1 beta density with the parameters $(\beta+p, \alpha)$. If the arbitrary function $f(\tilde{X}_2)$ is an arbitrary density then $\tilde{g_2}(\tilde{U_2})$ of (6.16) is a statistical density of $\tilde{U_2} = \tilde{X}_2^{\frac{1}{2}}\tilde{X}_1\tilde{X}_2^{\frac{1}{2}}$, where $\tilde{X}_1$ and $\tilde{X}_2$ are statistically independently distributed Hermitian positive definite $p \times p$ matrix random variables in the complex domain. In this case, from (6.16),

$$\frac{\tilde{\Gamma}_p(\beta+p)}{\tilde{\Gamma}_p(\alpha+\beta+p)}\tilde{g}_2(\tilde{U}_2) = \frac{|\det(\tilde{U}_2)|^{\beta}}{\tilde{\Gamma}_p(\alpha)}\int_{\tilde{V}>\tilde{U}_2>O}|\det(\tilde{V})|^{-\beta-\alpha}|\det(\tilde{V}-\tilde{U}_2)|^{\alpha-p}f(\tilde{V})\mathrm{d}\tilde{V} \tag{6.20}$$

$$= \tilde{K}^{-\alpha}_{2,\tilde{U}_2,\beta}f$$

where $\tilde{K}^{-\alpha}_{2,\tilde{U}_2,\beta}f$ is Erdélyi-Kober fractional integral of the second kind of order α and parameter β for the complex matrix-variate case. In the real matrix-variate case the right side of (6.20) will be of the following form:

$$\tilde{K}^{-\alpha}_{2,U_2,\beta}f = \frac{[\det(U_2)]^{\beta}}{\Gamma_p(\alpha)}\int_{V>U_2>O}[\det(V)]^{-\alpha-\beta}[\det(V-U_2)]^{\alpha-\frac{p+1}{2}}f(V)\mathrm{d}V.$$

These are called the Erdélyi-Kober fractional integrals because for $p=1$ they agree with Erdélyi-Kober fractional integral of the second kind. Note that for $p=1$ a Hermitian positive definite matrix is a real scalar positive variable.

Theorem 6.2 *For the Erdélyi-Kober fractional integral of the second kind of order α, as defined in* (6.20), *the M-transform is given by the following:*

$$\tilde{M}\{\tilde{K}^{-\alpha}_{2,\tilde{U}_2,\beta}f; s\} = \frac{\tilde{\Gamma}_p(\beta+s)}{\tilde{\Gamma}_p(\alpha+\beta+s)}\tilde{M}_f(s) \tag{6.21}$$

for $\Re(\beta+s) > p-1, \Re(\alpha) > p-1.$

Proof

$$\tilde{M}\{\tilde{K}_{2,\tilde{U_2},\beta}^{-\alpha}f;s\} = \int_{\tilde{U_2}>O} |\det(\tilde{U_2})|^{s-p}[\frac{|\det(\tilde{U_2})|^{\beta}}{\tilde{\Gamma}_p(\alpha)}$$

$$\times \int_{\tilde{V}>\tilde{U_2}>O} |\det(\tilde{V}-\tilde{U_2})|^{\alpha-p}|\det(\tilde{V})|^{-\alpha-\beta}f(\tilde{V})\mathrm{d}\tilde{V}]\mathrm{d}\tilde{U_2}.$$

Integrating out $\tilde{U_2}$ we have

$$\int_{O<\tilde{U_2}<\tilde{V}} |\det(\tilde{U_2})|^{\beta+s-p}|\det(\tilde{V}-\tilde{U_2})|^{\alpha-p}\mathrm{d}\tilde{U_2} = |\det(\tilde{V})|^{\alpha+\beta+s-p}\frac{\tilde{\Gamma}_p(\alpha)\tilde{\Gamma}_p(\beta+s)}{\tilde{\Gamma}_p(\alpha+\beta+s)}$$

for $\Re(\beta+s) > p-1, \Re(\alpha) > p-1$. Here the transformations are the following:

$$|\det(\tilde{V}-\tilde{U_2})|^{\alpha-p} = |\det(\tilde{V})|^{\alpha-p}|\det(I-\tilde{V}^{-\frac{1}{2}}\tilde{U_2}\tilde{V}^{-\frac{1}{2}})|^{\alpha-p}$$

$$\tilde{W} = \tilde{V}^{-\frac{1}{2}}\tilde{U_2}\tilde{V}^{-\frac{1}{2}} \Rightarrow \mathrm{d}\tilde{W} = |\det(\tilde{V})|^{-p}\mathrm{d}\tilde{U_2}.$$

Now by taking the integral over $\tilde{V}$ we have $\tilde{M}_f(s)$ and hence the result.

6.4.2 The Right-Sided Riemann-Liouville and Weyl Fractional Integrals in the Complex Matrix-Variate Case

Let $\phi_1(\tilde{X}_1) = 1$ and $\phi_2(\tilde{X}_2) = |\det(\tilde{X}_2)|^{\alpha}$. Then

$$f_1(\tilde{V}^{-\frac{1}{2}}\tilde{U_2}\tilde{V}^{-\frac{1}{2}})|\det(\tilde{V})|^{-p} = \frac{1}{\tilde{\Gamma}_p(\alpha)}|\det(\tilde{V})|^{-p}|\det(I-\tilde{V}^{-\frac{1}{2}}\tilde{U_2}\tilde{V}^{-\frac{1}{2}})|^{\alpha-p}$$

$$= \frac{1}{\tilde{\Gamma}_p(\alpha)}|\det(\tilde{V})|^{-\alpha}|\det(\tilde{V}-\tilde{U_2})|^{\alpha-p}.$$

Now we have

$$\tilde{g}_2(\tilde{U_2}) = \frac{1}{\tilde{\Gamma}_p(\alpha)}\int_{\tilde{V}>\tilde{U_2}>O} |\det(\tilde{V}-\tilde{U_2})|^{\alpha-p}f(\tilde{V})\mathrm{d}\tilde{V}$$

which is the right-sided or second kind Riemann-Liouville fractional integral in the complex matrix-variate case if $\tilde{V}$ is bounded above by a constant matrix $B = B^* > O$, and if not, it is the right-sided or second kind Weyl fractional integral for the complex matrix-variate case.

6.4.3 *Saigo and Related Fractional Integrals of the Second Kind*

Saigo fractional integrals in the real scalar case is defined in terms of a hypergeometric series of the type ${}_2F_1$ or Gauss' hypergeometric series. But ${}_2F_1$ has the defect that when it comes to Laplace transforms and Mellin transforms the series forms will have convergence problems and hence we will consider a general hypergeometric series in the complex matrix-variate case. For getting this special case from Definition 6.1 we will specialize $\phi_1(\tilde{X}_1)$ by taking a hypergeometric series for it. The definitions in the complex case will be slightly different from those in the real case, whether scalar case or matrix case. We need some notations, which will be introduced here. Let $K = (k_1, \ldots, k_p), k_1 + \ldots + k_p = k$, be the partitioning of a non-negative integer k into p parts, $k_1, \ldots, k_p$. Let

$$[\alpha]_K = \prod_{j=1}^{p} (\alpha - j + 1)_{k_j} = \frac{\tilde{\Gamma}_p(\alpha, K)}{\tilde{\Gamma}_p(\alpha)} \tag{6.22}$$

where $(\alpha)_m = \alpha(\alpha + 1)\ldots(\alpha + m - 1), (\alpha)_0 = 1, \alpha \neq 0$ is the Pochhammer symbol, and

$$\tilde{\Gamma}_p(\alpha, K) = \pi^{\frac{p(p-1)}{2}} \prod_{j=1}^{p} \Gamma(\alpha + k_j - j + 1)$$

$$= \tilde{\Gamma}_p(\alpha) \prod_{j=1}^{p} (\alpha - j + 1)_{k_j} = \tilde{\Gamma}_p(\alpha)[\alpha]_K. \tag{6.23}$$

A general hypergeometric series with m upper and n lower parameters and argument a complex matrix $\tilde{X}$ will be defined as the following series:

$${}_m\tilde{F}_n(\tilde{X}) = {}_m\tilde{F}_n(a_1, \ldots, a_m; b_1, \ldots, b_n; \tilde{X})$$

$$= \sum_{k=0}^{\infty} \sum_{K} \frac{[a_1]_K \ldots [a_m]_K}{[b_1]_K \ldots [b_n]_K} \frac{\tilde{C}_K(\tilde{X})}{k!} \tag{6.24}$$

where $\tilde{C}_K(\tilde{X})$ is the zonal polynomial of order k in the complex matrix-variate case. For the definition and details see Mathai [1, 5]. The series in (6.24) is convergent for all $\tilde{X}$ when $n \geq m$ and convergent for $\|\tilde{X}\| < 1$ when $m = n+1$ where $\|(\cdot)\|$ denotes a norm of $(\cdot)$. Since $f_1(\tilde{X}_1)$ has two factors $\phi_1(\tilde{X}_1)$ and $|\det(I - \tilde{X}_1)|^{\alpha - p}$ we can consider a hypergeometric series with argument $A\tilde{X}_1$ or with argument $A(I - \tilde{X}_1)$ where A is a real positive scalar constant or Hermitian positive definite constant matrix. For simplicity we may take $A = I$. Then the kernel of the integrals to be considered are of two types, we may call them I_A and I_B respectively, That is,

$$I_A = \int_{\tilde{V}} |\det(\tilde{V})|^{-p} \tilde{C}_K(\tilde{X}_1) |\det(I - \tilde{X}_1)|^{\alpha-p} f_2(\tilde{V}) \mathrm{d}\tilde{V} \tag{6.25}$$

and

$$I_B = \int_{\tilde{V}} |\det(\tilde{V})|^{-p} \tilde{C}_K(I - \tilde{X}_1) |\det(I - \tilde{X}_1)|^{\alpha-p} f_2(\tilde{V}) \mathrm{d}\tilde{V}. \tag{6.26}$$

For evaluating such integrals the following known results will be useful, which will be stated as lemmas.

Lemma 6.5 *For* $\tilde{Z}, \tilde{S}$ *Hermitian positive definite,* $\Re(\alpha) > p - 1$, $K = (k_1, \ldots, k_p), k_1 + \ldots + k_p = k$

$$\int_{\tilde{Z}>O} \mathrm{e}^{-\mathrm{tr}(\tilde{Z}\tilde{S})} |\det(\tilde{Z})|^{\alpha-p} \tilde{C}_K(\tilde{Z}\tilde{T}) \mathrm{d}\tilde{Z} = \tilde{\Gamma}_p(\alpha, K) |\det(\tilde{S})|^{-\alpha} \tilde{C}_K(\tilde{T}\tilde{S}^{-1}). \tag{6.27}$$

Lemma 6.6 *For* $O < \tilde{Z} < I, \Re(\alpha) > p - 1, \Re(\beta) > p - 1$

$$\int_{O<\tilde{Z}<I} |\det(\tilde{Z})|^{\alpha-p} |\det(I - \tilde{Z})|^{\beta-p} \tilde{C}_K(\tilde{Z}\tilde{S}) \mathrm{d}\tilde{Z} = \frac{\tilde{\Gamma}_p(\alpha, K) \tilde{\Gamma}_p(\beta) \tilde{C}_K(\tilde{S})}{\tilde{\Gamma}_p(\alpha + \beta, K)}. \tag{6.28}$$

When

$$\phi_1(\tilde{X}_1) = {}_m\tilde{F}_n(a_1, \ldots, a_m; b_1, \ldots, b_n; A\tilde{X}_1) \tag{6.29}$$

then the M-convolution of a product is given by the following:

$$\begin{aligned}\tilde{g}_2(\tilde{U}_2) = \frac{1}{\tilde{\Gamma}_p(\alpha)} \int_{\tilde{V}} |\det(\tilde{V})|^{-p} {}_m\tilde{F}_n(a_1, \ldots, a_m; b_1, \ldots, b_n; A\tilde{V}^{-\frac{1}{2}}\tilde{U}_2\tilde{V}^{-\frac{1}{2}}) \\ \times |\det(I - \tilde{V}^{-\frac{1}{2}}\tilde{U}_2\tilde{V}^{-\frac{1}{2}})|^{\alpha-p} \phi_2(\tilde{V}) f(\tilde{V}) \mathrm{d}\tilde{V}.\end{aligned}$$

The M-transform of the M-convolution of a product when $\phi_1(\tilde{X}_1)$ is given by (6.29) is given by the following theorem.

Theorem 6.3 *The M-transform, with parameter s, of the* $\tilde{g}_2(\tilde{U}_2)$ *of the M-convolution of a product when* $\phi_1(\tilde{X}_1)$ *is given by* (6.29) *is given by*

$$\tilde{M}\{\tilde{g}_2(\tilde{U}_2); s\} = \frac{\tilde{\Gamma}_p(s)}{\tilde{\Gamma}_p(s+\alpha)} {}_{m+1}\tilde{F}_{n+1}(a_1, \ldots, a_p, s; b_1, \ldots, b_n, s+\alpha; A)\tilde{M}_{f_2}(s) \tag{6.30}$$

for $\Re(s) > p-1, \Re(\alpha) > p-1$.

Proof The M-transform of $\tilde{g}_2(\tilde{U}_2)$ is available as

$$\tilde{M}\{f(\tilde{U}_2); s\} = \sum_{k=0}^{\infty}\sum_{K} \frac{[a_1]_K \ldots [a_m]_K}{[b_1]_K \ldots [b_n]_K k!} \int_{\tilde{U}_2 > O} |\det(\tilde{U}_2)|^{s-p}[\frac{1}{\tilde{\Gamma}_p(\alpha)}|\det(\tilde{V})|^{-p}$$
$$\times \tilde{C}_K(A\tilde{V}^{-\frac{1}{2}}\tilde{U}_2\tilde{V}^{-\frac{1}{2}})|\det(I-\tilde{V}^{-\frac{1}{2}}\tilde{U}\tilde{V}^{-\frac{1}{2}})|^{\alpha-p}\phi_2(\tilde{V})f(\tilde{V})\mathrm{d}\tilde{V}]\mathrm{d}\tilde{U}_2.$$

Put $\tilde{W} = \tilde{V}^{-\frac{1}{2}}\tilde{U}_2\tilde{V}^{-\frac{1}{2}}$. Then the $\tilde{W}$-integral is given by

$$\frac{1}{\tilde{\Gamma}_p(\alpha)} \int_{\tilde{W}} |\det(\tilde{W})|^{s-p}\tilde{C}_K(A\tilde{W})|\det(I-\tilde{W})|^{\alpha-p}\mathrm{d}\tilde{W}|\det(\tilde{V})|^{s-p}$$
$$= \frac{1}{\tilde{\Gamma}_p(\alpha)} \frac{\tilde{\Gamma}_p(s,K)\tilde{\Gamma}_p(\alpha)\tilde{C}_K(A)}{\tilde{\Gamma}_p(s+\alpha,K)}|\det(\tilde{V})|^{s-p},$$

from Lemma 6.6. Substituting these we have

$$\tilde{M}\{\tilde{g}_2(\tilde{U}_2); s\} = \frac{\tilde{\Gamma}_p(s)}{\tilde{\Gamma}_p(s+\alpha)} {}_{m+1}\tilde{F}_{n+1}(a_1, \ldots, a_m, s; b_1, \ldots, b_n, s+\alpha; A)\tilde{M}_{f_2}(s)$$

for $\Re(s) > p-1, \Re(\alpha) > p-1$.

A companion result can also be obtained by taking

$$\phi_1(\tilde{X}_1) = {}_m\tilde{F}_n(a_1, \ldots, a_m; b_1, \ldots, b_n; A(I-\tilde{X}_1)). \tag{6.31}$$

for $O < \tilde{X}_1 < I$ and $\phi_1(\tilde{X}_1) = 0$ elsewhere. Then the fractional integral of the second kind in the complex matrix-variate case will be of the following form:

$$\tilde{g}_2(\tilde{U}_2) = \sum_{k=0}^{\infty}\sum_{K} \frac{[a_1]_K \ldots [a_m]_K}{[b_1]_K \ldots [b_n]_K\, k!} \frac{1}{\tilde{\Gamma}_p(\alpha)} \int_{\tilde{V}} |\det(\tilde{V})|^{-p}\tilde{C}_K(A(I-\tilde{V}^{-\frac{1}{2}}\tilde{U}_2\tilde{V}^{-\frac{1}{2}}))$$
$$\times\, |\det(I-\tilde{V}^{-\frac{1}{2}}\tilde{U}\tilde{V}^{-\frac{1}{2}})|^{\alpha-p}\phi_2(\tilde{V})f(\tilde{V})\mathrm{d}\tilde{V}. \tag{6.32}$$

In this case the M-transform of $\tilde{g}_2(\tilde{U}_2)$ is given by the following theorem:

Theorem 6.4 *For the* $\phi_1(\tilde{X}_1)$ *defined in* (6.31) *the M-transform of the M-convolution of a product or M-transform of the fractional integral of the second kind in the complex matrix-variate case is given by*

$$\tilde{M}\{\tilde{g}_2(\tilde{U}_2); s\} = \frac{\tilde{\Gamma}_p(s)}{\tilde{\Gamma}_p(\alpha+s)}\tilde{M}_{f_2}(s)_{m+1}\tilde{F}_{n+1}(a_1,\ldots,a_m,\alpha;b_1,\ldots,b_n,\alpha+s;A) \tag{6.33}$$

for $\Re(\alpha) > p-1, \Re(s) > p-1.$

The proof is parallel to that in Theorem 6.3. Make the substitutions $\tilde{W} = \tilde{V}^{-\frac{1}{2}}\tilde{U}_2\tilde{V}^{-\frac{1}{2}}, \tilde{T} = I - \tilde{W}$. Then proceed as in the proof of Theorem 6.3 to establish the result.

Note 6.2 The Definition 6.1 covers all the known fractional integrals of the second kind or right-sided fractional integrals in the real or complex scalar and matrix cases. Hence Definition 6.1 can be taken as the definition for fractional integrals as Mellin convolution of a product in the scalar case and M-convolution of a product in the matrix-variate cases where the matrices are either real positive definite or Hermitian positive definite and the functions f_1 and f_2 are such that $f_2(\tilde{X}_2) = \phi_2(\tilde{X}_2)f(\tilde{X}_2)$ where ϕ_2 is a specified function and f is an arbitrary function, and

$$f_1(\tilde{X}_1) = \phi_1(\tilde{X}_1)\frac{1}{\tilde{\Gamma}_p(\alpha)}|\det(I - \tilde{X}_1)|^{\alpha-p} \tag{6.34}$$

for $O < \tilde{X}_1 < I$ and zero elsewhere, where $\phi_1(\tilde{X}_1)$ is a specified function, for $\Re(\alpha) > p-1$. In the real case the binomial factor is of the form

$$\frac{1}{\Gamma_p(\alpha)}[\det(I - X_1)]^{\alpha-\frac{p+1}{2}} \text{ for } O < X_1 < I,\ \Re(\alpha) > \frac{p-1}{2} \text{ and zero elsewhere.}$$

For $p = 1$ one gets the corresponding scalar versions.

Now, we will give a pathway generalized definition of fractional integrals of the second kind, which will encompass Definition 6.1 also.

6.4.4 A Pathway Generalized Definition of Fractional Integrals of the Second Kind in the Complex Matrix-Variate Case

It will be defined as in Definition 6.1 except that $f_1(\tilde{X}_1)$ will be of the following form:

$$f_1(\tilde{X}_1) = \phi_1(\tilde{X}_1)\frac{1}{\tilde{\Gamma}_p(\alpha)}|\det(I - a(1-q)\tilde{X}_1)|^{\alpha-p} \tag{6.35}$$

for $I - a(1-q)\tilde{X}_1 > O, a > 0, q < 1, \Re(\alpha) > p-1$ and zero elsewhere, where $\phi_1(\tilde{X}_1)$ is a specified function. Note that when $a = 1, q = 0$ this pathway extension agrees with Definition 6.1. Pathway idea may be seen from Mathai and Provost [4]. But for $-\infty < q < 1$, (6.35) describes a collection of functions or a path leading to the limiting form when $q \to 1_-$. Note that when $q < 1$ the functional form in (6.35) stays in the generalized type-1 beta family of functions. This extended definition also enables us to go to two other functional forms, namely for $q > 1$ (write $1 - q$ as $-(q - 1)$ for $q > 1$) we get into a wide class of functions belonging to type-2 beta family of functions and when $q \to 1$ we end up with a gamma family of functions. Thus, this pathway extended definition can be taken as a very general definition for fractional integrals of the second kind or the right-sided fractional integrals as the M-convolution of a product where one function f_1 is of the form in (6.35). Properties and M-transforms of the pathway extension can be studied parallel to the study in the non-extended situation and hence further details are omitted.

6.5 Fractional Integral of Order α and Parameter β of the First Kind in the Complex Matrix-variate Case

Here we will introduce a formal definition of fractional integral of the first kind in the complex matrix-variate case. In order to illustrate the process we will start with the evaluation of the density of a ratio. Let $\tilde{X}_1$ and $\tilde{X}_2$ be $p \times p$ Hermitian positive definite matrix-variate random variables in the complex domain and statistically independently distributed with the density functions $f_1(\tilde{X}_1)$ and $f_2(\tilde{X}_2)$ respectively, where f_1 and f_2 are real-valued scalar functions of the matrices $\tilde{X}_1$ and $\tilde{X}_2$ respectively. Let us consider the following ratio $\tilde{U_1} = \tilde{X}_2^{\frac{1}{2}}\tilde{X}_1^{-1}\tilde{X}_2^{\frac{1}{2}}$. Consider the transformation $\tilde{X}_2 = \tilde{V}$ and $\tilde{X}_1 = \tilde{V}^{\frac{1}{2}}\tilde{U_1}^{-1}\tilde{V}^{\frac{1}{2}}$. The Jacobinan is given by

$$\mathrm{d}\tilde{X}_1 \wedge \mathrm{d}\tilde{X}_2 = |\det(V)|^p |\det(\tilde{U_1}\tilde{U_1}^*)|^{-p} \mathrm{d}\tilde{U_1} \wedge \mathrm{d}\tilde{V} \tag{6.36}$$

from Lemmas 6.2 and 6.4. Let the density of $\tilde{U_1}$ be denoted by $\tilde{g}_1(\tilde{U_1})$. Then the density of $\tilde{U_1}$ is given by

$$\tilde{g}_1(\tilde{U_1}) = \int_{\tilde{V}} |\det(\tilde{V})|^p |\det(\tilde{U_1}\tilde{U_1}^*)|^{-p} f_1(\tilde{V}^{\frac{1}{2}}\tilde{U_1}^{-1}\tilde{V}^{\frac{1}{2}}) f_2(\tilde{V}) \mathrm{d}\tilde{V}. \tag{6.37}$$

If f_1 and f_2 are arbitrary real-valued scalar functions of $p \times p$ Hermitian positive definite matrices in the complex domain, need not be densities, and if (6.37) is convergent, then (6.37) is called the M-convolution of a ratio. Fractional integral of the first kind of order α in the complex matrix-variate case will be defined in terms of M-convolution of a ratio.

Definition 6.2 (Fractional Integral of Order α of the First Kind in the Complex Matrix-variate Case) It is defined as $\tilde{g}_1(\tilde{U}_1)$ of (6.37) where $f_2(\tilde{X}_2) = \phi_2(\tilde{X}_2)f(\tilde{X}_2)$ where ϕ_2 is a specified function and f is an arbitrary function, and $f_1(\tilde{X}_1) = \phi_1(\tilde{X}_1)\frac{|\det(I-\tilde{X}_1)|^{\alpha-p}}{\tilde{\Gamma}_p(\alpha)}$, $\Re(\alpha) > p-1$ where ϕ_1 is a specified function.

In this case, $\tilde{g}_1(\tilde{U}_1)$ of (6.37) becomes

$$\begin{aligned}\tilde{g}_1(\tilde{U}_1) &= \frac{1}{\tilde{\Gamma}_p(\alpha)}\int_{\tilde{V}} |\det(I - \tilde{V}^{\frac{1}{2}}\tilde{U}_1^{-1}\tilde{V}^{\frac{1}{2}})|^{\alpha-p}\phi_1(\tilde{V}^{\frac{1}{2}}\tilde{U}_1^{-1}\tilde{V}^{\frac{1}{2}})\\ &\quad\times |\det(\tilde{V})|^{p}|\det(\tilde{U}_1\tilde{U}_1^{*})|^{-p}\phi_2(\tilde{V})f(\tilde{V})\mathrm{d}\tilde{V}\\ &= \frac{1}{\tilde{\Gamma}_p(\alpha)}\int_{O<\tilde{V}<\tilde{U}_1} \phi_1(\tilde{V}^{\frac{1}{2}}\tilde{U}_1^{-1}\tilde{V}^{\frac{1}{2}})|\det(\tilde{V})|^{p}|\det(\tilde{U}_1)|^{-\alpha-p}\\ &\quad\times |\det(\tilde{U}_1 - \tilde{V})|^{\alpha-p}\phi_2(\tilde{V})f(\tilde{V})\mathrm{d}\tilde{V}. \end{aligned}\tag{6.38}$$

In the real case (6.38) will be the following:

$$\begin{aligned}g_1(U_1) &= \frac{1}{\Gamma_p(\alpha)}\int_{O<V<U_1} \phi_1(V^{\frac{1}{2}}U_1^{-1}V^{\frac{1}{2}})[\det(V)]^{\frac{p+1}{2}}[\det(U_1)]^{-\alpha-\frac{p+1}{2}}\\ &\quad\times [\det(U_1 - V)]^{\alpha-\frac{p+1}{2}}\phi_2(V)f(V)\mathrm{d}V. \end{aligned}\tag{6.39}$$

Theorem 6.5 *If $f_1(\tilde{X}_1)$ is a type-1 beta density with parameters (β,α) and if $f_2(\tilde{X}_2) = f(\tilde{X}_2)$ is an arbitrary density then ϕ_1 of* (6.37) *is given by*

$$\phi_1(\tilde{X}_1) = \frac{\tilde{\Gamma}_p(\alpha+\beta)}{\tilde{\Gamma}_p(\beta)}|\det(\tilde{X}_1)|^{\beta-p}\tag{6.40}$$

and the density of the ratio, $\tilde{g}_1(\tilde{U}_1)$, is given by

$$\tilde{g}_1(\tilde{U}_1)=\frac{\tilde{\Gamma}_p(\alpha+\beta)}{\tilde{\Gamma}_p(\beta)}\frac{|\det(\tilde{U}_1)|^{-\alpha-\beta}}{\tilde{\Gamma}_p(\alpha)}\int_{O<\tilde{V}<\tilde{U}_1}|\det(\tilde{V})|^{\beta}|\det(\tilde{U}_1-\tilde{V})|^{\alpha-p}f(\tilde{V})\mathrm{d}\tilde{V}.\tag{6.41}$$

Note that the result is available from (6.39) by substituting for ϕ_1 from (6.40).

By rewriting (6.41) we can define Erdélyi-Kober fractional integral of the first kind of order α in the complex matrix-variate case.

6.5.1 Erdélyi-Kober Fractional Integral of Order α of the First Kind for Complex Matrix-Variate Case

It will be denoted by $\tilde{K}^{-\alpha}_{1,\tilde{U}_1,\beta} f$, and it is given by the following:

$$\frac{\tilde{\Gamma}_p(\beta)}{\tilde{\Gamma}_p(\alpha+\beta)}\tilde{g}_1(\tilde{U}_1) = \tilde{K}^{-\alpha}_{1,\tilde{U}_1,\beta} f$$

where

$$\tilde{K}^{-\alpha}_{1,\tilde{U}_1,\beta} f = \frac{|\det(\tilde{U}_1)|^{-\alpha-\beta}}{\tilde{\Gamma}_p(\alpha)} \int_{O<\tilde{V}<\tilde{U}_1} |\det(\tilde{V})|^{\beta} |\det(\tilde{U}_1 - \tilde{V})|^{\alpha-p} f(\tilde{V}) \mathrm{d}\tilde{V}. \tag{6.42}$$

Note that $\tilde{g}_1(\tilde{U}_1)$ in (6.41) and (6.42) is the statistical density of a ratio of the form $\tilde{U}_1 = \tilde{X}_2^{\frac{1}{2}} \tilde{X}_1^{-1} \tilde{X}_2^{\frac{1}{2}}$, where $\tilde{X}_1 = \tilde{X}_1^* > O$ and $\tilde{X}_2 = \tilde{X}_2^* > O$ are $p \times p$ statistically independently distributed Hermitian positive definite matrix random variables where $\tilde{X}_1$ has a type-1 beta density with parameters (β, α) and $\tilde{X}_2$ has an arbitrary density $f(\tilde{X}_2)$.

Theorem 6.6 *The M-transform of the Erdélyi-Kober fractional integral of the first kind as defined in* (6.42) *is given by*

$$\tilde{M}\{\tilde{K}^{-\alpha}_{1,\tilde{U}_1,\beta} f; s\} = \frac{\tilde{\Gamma}_p(\beta+p-s)}{\tilde{\Gamma}_p(\alpha+\beta+p-s)} \tilde{M}_f(s)$$

for $\Re(\alpha) > p-1, \Re(\beta+p-s) > p-1$.

Proof

$$\tilde{M}\{\tilde{K}^{-\alpha}_{1,\tilde{U}_1,\beta} f; s\} = \frac{1}{\tilde{\Gamma}_p(\alpha)} \int_{\tilde{U}_1 > O} |\det(\tilde{U}_1)|^{s-p-\alpha-\beta}$$
$$\times \Big[\int_{O<\tilde{V}<\tilde{U}_1} |\det(\tilde{V})|^{\beta} |\det(\tilde{U}_1 - \tilde{V})|^{\alpha-p} f(\tilde{V}) \mathrm{d}\tilde{V}\Big] \mathrm{d}\tilde{U}_1.$$

Put $\tilde{W} = \tilde{U}_1 - \tilde{V}$. Then the $\tilde{U}_1$-integral is given by

$$\int_{\tilde{U}_1>\tilde{V}>O} |\det(\tilde{U}_1)|^{-\alpha-\beta+s-p} |\det(\tilde{U}_1 - \tilde{V})|^{\alpha-p} \mathrm{d}\tilde{U}_1$$
$$= \int_{\tilde{W}>O} |\det(\tilde{V}+\tilde{W})|^{-\alpha-\beta+s-p} |\det(\tilde{W})|^{\alpha-p} \mathrm{d}\tilde{W}$$

$$= |\det(\tilde{V})|^{-\alpha-\beta+s-p} \int_{\tilde{W}>O} |\det(I + \tilde{V}^{-\frac{1}{2}}\tilde{W}\tilde{V}^{-\frac{1}{2}})|^{-\alpha-\beta+s-p} |\det(\tilde{W})|^{\alpha-p} \mathrm{d}\tilde{W}$$

$$= |\det(\tilde{V})|^{-\beta+s-p} \int_{\tilde{T}>O} |\det(\tilde{T})|^{\alpha-p} |\det(I + \tilde{T})|^{-\alpha-\beta+s-p} \mathrm{d}\tilde{T}, \tilde{T} = \tilde{V}^{-\frac{1}{2}}\tilde{W}\tilde{V}^{-\frac{1}{2}}$$

$$= |\det(\tilde{V})|^{-\beta+s-p} \tilde{\Gamma}_p(\alpha) \frac{\tilde{\Gamma}_p(\beta + p - s)}{\tilde{\Gamma}_p(\alpha + \beta + p - s)}$$

for $\Re(\alpha) > p - 1, \Re(\beta + p - s) > p - 1$. The integral above is evaluated by using a complex matrix-variate type-2 beta integral and the substitution is $\tilde{T} = \tilde{V}^{-\frac{1}{2}}\tilde{W}\tilde{V}^{-\frac{1}{2}}$. Now, evaluating the $\tilde{V}$-integral we have the M-transform of f and hence the result.

6.5.2 *Riemann-Liouvile and Weyl Fractional Integrals of the First Kind of Order α for the Complex Matrix-Variate Case*

In Definition 6.2, put

$$\phi_1(\tilde{X}_1) = |\det(\tilde{X}_1)|^{-\alpha-p} \text{ and } \phi_2(\tilde{X}_2) = |\det(\tilde{X}_2)|^{\alpha}.$$

Then $\tilde{g}_1(\tilde{U}_1)$ of (6.38) will reduce to the Riemmann-Liouville left-sided fractional integral of order α for the complex matrix variate case, denoted by $\tilde{D}^{-\alpha}_{1,\tilde{U}_1} f$, and given by

$$\tilde{D}^{-\alpha}_{1,\tilde{U}_1} f = \frac{1}{\tilde{\Gamma}_p(\alpha)} \int_{\tilde{V}<\tilde{U}_1} |\det(\tilde{U}_1 - \tilde{V})|^{\alpha-p} f(\tilde{V}) \mathrm{d}\tilde{V}. \tag{6.43}$$

If $\tilde{V}$ is bounded below by a constant matrix A then it is the Riemann-Liouville left-sided fractional integral and otherwise it is the left-sided Weyl fractional integral.

In Definition 6.2, we can replace $\phi_1(\tilde{X}_1)$ by any specified function of $\tilde{X}_1$ and then consider the M-convolution of a ratio. Then we get a corresponding fractional integral operator of the first kind in the complex matrix-variate case. For example, if ϕ_1 is a general hypergeometric series with argument either $A\tilde{X}_1$ or $A(I - \tilde{X}_1)$ for $O < \tilde{X}_1 < I$ and zero elsewhere then we can obtain the extension to complex matrix-variate case of Saigo and related fractional integral operators. The procedure is parallel to that in Sect. 6.4.3. One can consider a pathway extension of $f_1(\tilde{X}_1)$ as in Eq. (6.35) and then consider the M-convolution of a ratio. Then we have a pathway extension of fractional integrals of the first kind. Since the steps will be parallel we will not elaborate these items here.

Fractional integrals involving many matrix variables, in the real and complex cases, can be dealt with without much difficulty. The real case is discussed in Chap. 5. Results parallel to those in many real scalar variables in Chap. 4 and the real many matrices case in Chap. 5 can be obtained for the many complex matrix-variates case. In order to limit the size of this monograph, those materials will not be included here.

References

1. A.M. Mathai, *Jacobians of Matrix Transformations and Functions of Matrix Argument* (World Scientific Publishing, New York, 1997)
2. A.M. Mathai, Fractional integral operators in the complex matrix-variate case. Linear Algebra Appl. **439**, 2901–2913 (2013)
3. A.M. Mathai, Evaluation of matrix-variate gamma and beta integrals. Appl. Math. Comput. **247**, 312–318 (2014)
4. A.M. Mathai, S.B. Provost, Some complex matrix-variate statistical distributions on rectangular matrices. Linear Algebra Appl. **410**, 198–216 (2005)
5. A.M. Mathai, S.B. Provost, T. Hayakawa, *Bilinear Forms and Zonal Polynomials*. Lecture Notes in Statistics Series, vol. 102 (Springer, New York, 1995)

Index

A. M. Mathai, H. J. Haubold, *Erdélyi–Kober Fractional Calculus*, SpringerBriefs in Mathematical Physics 31, https://doi.org/10.1007/978-981-13-1159-8

Printed in the United States
By Bookmasters